Layout Techniques for MOSFETs

Synthesis Lectures on Emerging Engineering Technology

Kris Iniewski, *Redlen Technologies, Inc.*

Layout Techniques for MOSFETS
Salvador Pinillos Gimenez
March 2016

Resistive Random Access Memory (RRAM): From Devices to Array Architectures
Shimeng Yu
March 2016

The Digital Revolution
Bob Merritt
February 2016

Compound Semiconductor Materials and Devices
Zhaojun Liu, Tongde Huang, Qiang Li, Xing Lu, Xinbo Zou
February 2016

New Prospects of Integrating Low Substrate Temperatures with Scaling-Sustained Device Architectural Innovation
Nabil Shovon Ashraf, Shawon Alam, Mohaiminul Alam
February 2015

Advances in Reflectometric Sensing for Industrial Applications
Andrea Cataldo, Egidio De Benedetto, and Giuseppe Cannazza
January 2016

Sustaining Moore's Law: Uncertainty Leading to a Certainty of IoT Revolution
Apek Mulay
September 2015

Layout Techniques for MOSFETS
Salvador Pinillos Gimenez

ISBN: 978-3-031-00903-7   print
ISBN: 978-3-031-02031-5   ebook

DOI 10.1007/978-3-031-02031-5

A Publication in the Springer series
SYNTHESIS LECTURES ON EMERGING ENGINEERING TECHNOLOGIES, #7
Series Editors: Kris Iniewski, Redlen Technologies, Inc.

Series ISSN 2381-1412 Print   2381-1439 Electronic

Layout Techniques for MOSFETS

Salvador Pinillos Gimenez
FEI University Center, Brazil

SYNTHESIS LECTURES ON EMERGING ENGINEERING TECHNOLOGIES
#7

ABSTRACT

This book aims at describing in detail the different layout techniques for remarkably boosting the electrical performance and the ionizing radiation tolerance of planar Metal-Oxide-Semiconductor (MOS) Field Effect Transistors (MOSFETs), without adding any costs to the current planar Complementary MOS (CMOS) integrated circuits (ICs) manufacturing processes. These innovative layout styles are based on pn junctions engineering between the drain/source and channel regions or simply MOSFET gate layout change. These interesting layout structures are capable of incorporating new effects in the MOSFET structures, such as the Longitudinal Corner Effect (LCE), the Parallel connection of MOSFETs with Different Channel Lengths Effect (PAMDLE), the Deactivation of the Parallel MOSFETs in the Bird's Beak Regions (DEPAMBBRE), and the Drain Leakage Current Reduction Effect (DLECRE), which are still seldom explored by the semiconductor and CMOS ICs industries. Several three-dimensional (3D) numerical simulations and experimental works are referenced in this book to show how these layout techniques can help the designers to reach the analog and digital CMOS ICs specifications with no additional cost. Furthermore, the electrical performance and ionizing radiation robustness of the analog and digital CMOS ICs can significantly be increased by using this gate layout approach.

KEYWORDS

layout techniques, Circular Annular MOSFET, pillar surrounding gate MOSFET, Cynthia MOSFET, Diamond MOSFET, Octo MOSFET, Ellipsoidal MOSFET, Fish MOSFET, Wave MOSFET, LCE, PAMDLE, DEPAMBBRE, DLEFRE, high temperature, Ionizing Radiation Effects, TID and SEE

Dedication

I dedicate this book to my family and my parents (in memoriam).

I would also like to dedicate this work to the FEI University Center and my students (scientific initiation, master's degree, and Ph.D.), who have supported me in my research.

Contents

Acknowledgments

First and foremost, I would like to thank God!

I want to thank my family for supporting me during the writing of this book, and Professor Krzysztof (Kris) Iniewskiis for inviting me to write this book. It has been an honor.

I would like to show my gratitude to the FEI University Center, all researchers, professors, and students who supported me in my studies.

My sincere thanks also go to Professor Dr. Denis Flandre and Mr. Christian Renaux from UCL for fabricating these new devices and their technical discussions. I also want to thank Prof. Dr. Cor Claeys and Prof. Dr. Eddy Simon from imec for their technical support.

I am also grateful to the MOSIS team, from the MOSIS Educational Program (MEP), for manufacturing these new devices.

Last but not least, I thank my colleagues from the Integrated Electronics Devices (IED) group of Electrical Engineering Department at the FEI University Center.

I also thank Victor A. Santos for revising the English version.

CHAPTER 1

Introduction

Specialists have been making a lot of effort in research and development (R&D) in the last 57 years (after the implementation of the first integrated circuit (IC) designed by Jack St. Claire Kilby, in 1958 at Texas Instruments [1]). Their main objectives were to reduce dimensions and boost the electrical performance of the Metal-Oxide-Semiconductor (MOS) Field Effect Transistor (MOS-FET), and consequently their integrated circuits (ICs) [2]. In 1965, Gordon Moore proposed the Moore's Law, regarding the dimensional reduction of MOSFETs over time [3], which has been valid until now. Due to all these efforts, the MOSFETs dimensions have evolved from millimetric to nanometric scales (six orders of magnitude in approximately 60 years, in other words, almost one order of magnitude of MOSFETs dimensional reduction per decade).

For all these years, the R&D in the micro/nanoelectronics has basically been classified in three main categories: new structures (double gate [4], FinFET [5, 6], MuGFET [7], Gate-All-Around [8], UTB [9], UTBB [10, 11], Junctionless MOSFET [12, 13], TFET [14], etc.), new materials (Silicon-on-Insulator, SOI [15], Silicon-on-Sapphire, SOS [16], High-k material [17], Germanium [18], Silicon-Germanium [19], etc.), and new manufacturing processes to implement MOSFETs (plasma [20], photolithograph [20, 21], etc.).

However, there is another category, still unexplored by the semiconductor and ICs companies, that regards the "PN Junctions Engineering between the Drain/Source and the Channel Regions," or simply a gate layout change of the MOSFETs, i.e., from a rectangular to a non-standard geometries (circular, ellipsoidal, hexagonal, octagonal, etc.) [22–54].

The non-standard (non-rectangular) gate layout styles for MOSFETs are capable of including other unknown effects in the MOSFET structure, which can be explored to boost their electrical performance and ionizing radiation tolerance. The longitudinal corner effect (LCE), parallel connections of MOSFETs with different channel lengths effect (PAMDLE), deactivation the parasitic MOSFETs in the bird's beak regions effect (DEPAMBBRE), and Drain Leakage Current Reduction Effect (DLECRE) [22–54] are some examples of these new effects in MOS-FETs when we change their gate layouts. The main feature of this layout technique is that it does not add any extra costs to the current planar complementary MOS (CMOS) ICs manufacturing processes [22–54].

Some examples of these innovative layout styles for MOSFETs are: annular/circular gate (A/C-G) [22–24], overlapping-circular gate (O-CG) [25, 26], Wave ("S" gate shape) [27–29], hexagonal gate (Diamond) [30–40], multiple edges gate (M-E) [41], octagonal gate (Octo) [41–49], Ellipsoidal [50–52], and Fish ("<" shaped gate) [53, 54].

In this context, the aim of this book is to describe the origin of this new field of research in micro/nanoelectronics, still unexplored by the semiconductor and CMOS ICs industries. This new research mainly focuses on using non-standard innovative layout styles for planar MOSFETs to boost their resultant longitudinal electric field (LEF) along the channel length (L). As a result, it is capable of increasing the drift velocity of their mobile charge carriers along the L and consequently boosting their drain current (IDS) in relation to the standard rectangular transistors counterparts, regarding the same gate area (AG) and bias conditions (LCE) [22–54]. Besides, the new effects caused by the use of these non-standard gate shapes in MOSFETs structures (PAMDLE, DEPAMBBRE, and DLECRE) are also studied, modeled, qualified, and quantified in this book [22–54].

Furthermore, this book also aims at describing in detail the different innovative layout techniques, encouraging their use in order to remarkably improve the MOSFETs electrical performance, regarding any planar CMOS ICs technologies.

Therefore, taking into account all the efforts that have been made so far to extend the planar CMOS ICs technologies (new manufacturing processes, materials, and new planar MOSFET structures), with these innovative layout techniques, we will certainly be able to further boost the electrical performance of the transistors remarkably in order to meet the never-ending demand for efficiency and low cost.

PROBLEMS AND QUESTIONS

1. On average, regarding the history of micro/nanoelectronics, what has the dimensional reduction index of MOFETs per decade been so far?

2. How are the main R&D in the micro/nanoelectronics area classified today?

3. Is there another way to improve the electrical performance of MOSFETs regarding R&D in the micro/nanoelectronics area, still unexplored by the semiconductor and planar CMOS ICs industries?

4. What are the new effects incorporated in the MOSFETs structures when we use non-standard gate geometries (PN Junctions Engineering between the Drain/Source and the Channel Regions)?

5. Give some examples of non-standard gate geometries to implement MOSFETs that are capable of improving their electrical performance and ionizing radiation tolerance?

6. What is your opinion if we combine these innovative layout techniques with the current and sophisticated planar CMOS ICs technologies (double gate, UTB, UTBB, TFET, etc.)?

CHAPTER 2

The Origin of the Innovative Layout Techniques for MOSFETs

Initially, this chapter describes how the Corner Effect (CE) was explored to improve the electrical performance and the ionizing radiation tolerance of the planar MOSFETs. This effect occurs in the multiple gates MOSFETs, i.e., in the MuGFETs with three or more gates. The CE is responsible for increasing the resultant electric field in the regions near to the vertices of the junctions of two different gates regions in MuGFETs, due to the gate voltage. This effect is responsible for reducing the threshold voltage (V_{TH}) of MuGFETs in these gate regions and consequently it reduces the electrostatic controllability of the transistors gate (undesired effect). Based on the CE concept that happens in the MuGFETs, the author applied it in the longitudinal direction of the MOSFETs channel length in order to increase the resultant longitudinal electric field due to the drain voltage (V_{DS}). As a consequence of this action, we could observe that the drift velocity of the mobile charge carriers in the transistor channel increase, the drain current (I_{DS}) and their corresponding analog and digital parameters and figures of merit could be boosted controllably.

Furthermore, some innovative layout styles for planar MOSFETs are also introduced and the new effects associated with them are discussed and modeled. The main results of three-dimensional (3D) numerical simulations and experimental data are reported in order to demonstrate that these new layout techniques are capable of boosting the MOSFETs electrical performance and the ionizing radiation tolerance substantially, without adding any extra cost to the current and sophisticated CMOS ICs manufacturing processes.

2.1 OBSERVING AND COMBINING DIFFERENT NEW EFFECTS IN MOSFETS

In the standard (rectangular-section) Gate-All-Around (GAA) MOSFETs [55] or Pillar (square-section) Surrounding Gate SOI MOSFETs [56], two perpendicular electric fields (PEF) vector components ($\vec{\varepsilon_1}$ and $\vec{\varepsilon_2}$) appear in the regions near to the devices gate corners due to the gate electrode material covering the silicon film of the channel region, regarding a bias from the gate to the source (V_{GS}). On the other hand, there is only one PEF vector component ($\varepsilon_1 = \vec{\varepsilon_2}$) in the rest of the channel outside the regions near the corners. At the corners of the silicon film of the channel region, the resultant PEF are higher (vector sum: $\varepsilon_T = \vec{\varepsilon_1} + \vec{\varepsilon_2}$) than those found in the rest of the channel, as illustrated in Figure 2.1. This effect was named the Corner effect (CE) [55] and it is responsible for reducing the threshold voltage (V_{TH}) in the corners of the device channel region. Therefore, for a specific V_{GS} higher than V_{TH}, the drain current (I_{DS}) further flows in the corners compared to other parts of the channel region [55, 56].

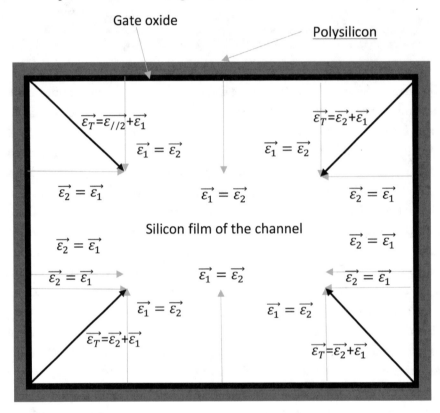

Figure 2.1: The PEF vector components ($\vec{\varepsilon_1}$ and $\vec{\varepsilon_2}$) and their resultants due to the VGS, where the CE happens.

Therefore, the CE is an undesired effect in 3D devices, which tends to reduce the electrostatic controllability by the V_{GS}, but it is capable of boosting the resultant PEF in regions that present corners [55, 56]. So, this effect was the first important piece of information used by the author to propose the innovative layout styles for MOSFETs to boost their electrical performance.

Furthermore, by investigating the most popular non-standard circular/annular gate MOS-FETs (CAGM), we observe that their I_{DS} is higher when operating in internal drain bias configuration (IDBC) than in external drain bias configuration (EDBC), (Figure 2.2) [22].

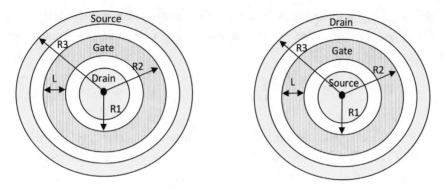

Figure 2.2: Simplified layout of the same CAGM, which is operating in IDBC (a) and EDBC (b), respectively.

In Figure 2.2, L is the channel length, which is equal to "R2-R1," where R1 and R2 are the internal and external radiuses that define the CAGM channel length, respectively, and R3 is the radius that defines the device's external edges [22].

These different behaviors of the CAGM I_{DS} occur because they depend on the drain bias configuration type, where the longitudinal electric field (LEF) along the channel region varies as a consequence of the differences between the number of LEF lines per pn junction area (A_{PNJ}) between the drain and the silicon film of the channel regions, and the silicon film of the channel and the source regions. This can obviously be justified as a result of the differences between the drain and source A_{PNJ}, according to Figure 2.2. Quite simply, the LEF densities are different in the channel regions near the drain and source regions, regarding a drain to source voltage (V_{DS}). These LEF densities depend on the type of the drain bias configurations (IDBC and EDBC). If we consider that the CAGM is operating in IDBC, the LEF is smaller in the channel region near the source region and is higher near the drain region because the A_{PNJ} between the source and the silicon film regions of the channel is higher than the one between the silicon film of the channel and the drain regions. The same consideration can be made regarding the CAGM operating in the EDBC, i.e., the LEF is smaller in the channel region near the drain and higher near the source. This happens due to the A_{PNJ} being higher between the drain and the silicon film regions of the channel than the A_{PNJ} between the silicon film of the channel region and the source [22].

Therefore, the circular/annular layout style for MOSFET is capable of increasing the LEF along the channel length on account of the asymmetric gate geometry [22]. This was the second remarkable piece of information used by the author to design the innovative layout styles for MOS-FETs to enhance their electrical performance and ionizing radiation tolerance.

Based on these two studies performed and reported previously, it was possible to combine the CE in the longitudinal direction of the channel length, called Longitudinal Corner Effect (LCE), to further increase the resultant LEF along the channel of the MOSFETs. This was possible because the MOSFET I_{DS} is directly proportional to the product of the inversion layer charge (Q_{inv}) induced by V_{GS} and the drift velocity of the mobile charge carriers in the channel, according to Equation (2.1) [57, 58]:

$$\overrightarrow{\vartheta}_{//} = \mu_i . \overrightarrow{\varepsilon}_{//} \tag{2.1},$$

where μ_i is the mobile charge carriers mobility in the channel (i=p, if p-type MOSFET and i=n if n-type MOSFET) and $\overrightarrow{\varepsilon}_{//}$ is the resultant LEF along the channel length due to the V_{DS} [57, 58].

In this context, a first non-typical layout style was specially proposed to boost the electrical performance and the ionizing radiation robustness of the MOSFETs, which have a hexagonal gate shape named "Diamond" [22–54]. Subsequently, the octagonal gate MOSFET (Octo MOSFET, OM) was carefully suggested to further enhance the resultant LEF along the channel, the Electrostatic-Discharge (ESD) tolerance, and the breakdown voltage (BV_{DS}) in relation to the Diamond MOSFET (DM) [41–49]. Other innovative layout styles were also personally designed regarding different planar CMOS ICs applications, such as Wave [27–29], Ellipsoidal [49–51], and Fish [52, 53], which will be discussed in detail in the next chapters of this book.

PROBLEMS AND QUESTIONS

1. What is the corner effect (CE) in 3D MOSFETs, such as the standard Gate-All-Around (GAA) MOSFETs with rectangular-section or Pillar (square-section) Surrounding Gate SOI MOSFETs? What does the corner effect cause in these devices?

2. What is the most popular non-standard gate layout style currently used in micro/nano-electronics?

3. Is the CAGM geometrically symmetric? Justify.

4. How can CAGM be configured in terms of the drain bias?

5. Does the CAGM electrical behavior change depending on the type of drain bias configuration? Justify.

6. Is the CAGM longitudinal electric field (LEF) constant along its channel length? Justify.

7. How was the diamond layout style for MOSFET invented?

CHAPTER 3

Diamond MOSFET (Hexagonal Gate Geometry)

This innovative layout style for MOSFETs was carefully designed in order to use the corner effect (CE) in the longitudinal direction of the channel, named Longitudinal Corner Effect (LCE), to boost its resultant longitudinal electric field ($\overrightarrow{\varepsilon_{//}}$). Consequently, the drift velocity of the mobile charge carriers ($\overrightarrow{\vartheta_{//}}$) also increases, according to Equation (3.1) [56]

$$\overrightarrow{\vartheta_{//}} = \mu_i . \overrightarrow{\varepsilon_{//}} \tag{3.1},$$

where i can be respectively n or p, depending on the n or p types of the mobile charge carriers of the MOSFETs.

As the MOSFET drain current (I_{DS}) is given by the equation (3.2) [57, 58]

$$\overrightarrow{I_{DS}} = Q_{inv} . \overrightarrow{\vartheta_{//}} = Q_{inv} . \mu_i . \overrightarrow{\varepsilon_{//}} \tag{3.2},$$

where Q_{inv} is the inversion mobile charge in the channel due to the voltage between the gate to source (V_{GS}), it is possible to boost the I_{DS} magnitude and therefore the analog and digital MOSFET electrical performance by improving the $\overrightarrow{\varepsilon_{//}}$ through the LCE.

This can be achieved by changing the MOSFET gate geometry from rectangular to hexagonal, according to Figure 3.1, which illustrates the simplified diamond layout style for MOSFET, indicating the LEF vector components ($\overrightarrow{\varepsilon_{//1}}$ and $\overrightarrow{\varepsilon_{//2}}$) in the point Q and the corresponding resultant LEF vector ($\overrightarrow{\varepsilon_{//DM}} = \overrightarrow{\varepsilon_{//1}} + \overrightarrow{\varepsilon_{//2}}$) due to the drain to source voltage (V_{DS}).

In Figure 3.1, b and B are the smallest and largest channel lengths, respectively, W is the MOSFET channel width and α is the angle composed of the triangular area of the hexagonal shape.

Observe that the LEF vector components are perpendicular to the pn metallurgical junctions defined between the drain/channel and channel/source regions and therefore the hexagonal gate shape is capable of generating two LEF vectorial components with the same magnitude ($\overrightarrow{\varepsilon_{//1}} + \overrightarrow{\varepsilon_{//2}}$ = $\overrightarrow{\varepsilon_{//}}$), instead of a single one, as it occurs in the typical rectangular gate geometry for MOSFETs. Therefore, the resultant LEF ($\overrightarrow{\varepsilon_{//\_DM}}$) along the channel length (L) of the Diamond MOSFET (DM) is higher than the one found in the standard rectangular MOSFET (RM). As a consequence of the LCE effect, the electrical performance of the DM is higher than the one observed in the RM, regarding the same gate area (A_G) and bias conditions, which can be controlled by the α angle of the hexagonal shape gate. Thus, we can obtain a higher DM electrical performance by reducing the value of the α angle of the hexagonal gate shape, because the LEF vectorial sum is given by $\sqrt{\varepsilon_{//1}^2 + \varepsilon_{//2}^2 + 2.\varepsilon_{//1}.\varepsilon_{//2}.\cos(\alpha)}$, which depends on the $\cos(\alpha)$. Therefore, for lower α angles, the resultant LEF is higher in Diamond MOSFET. It is relevant to highlight that the α angle is limited by the planar CMOS ICs manufacturing processes [30–40].

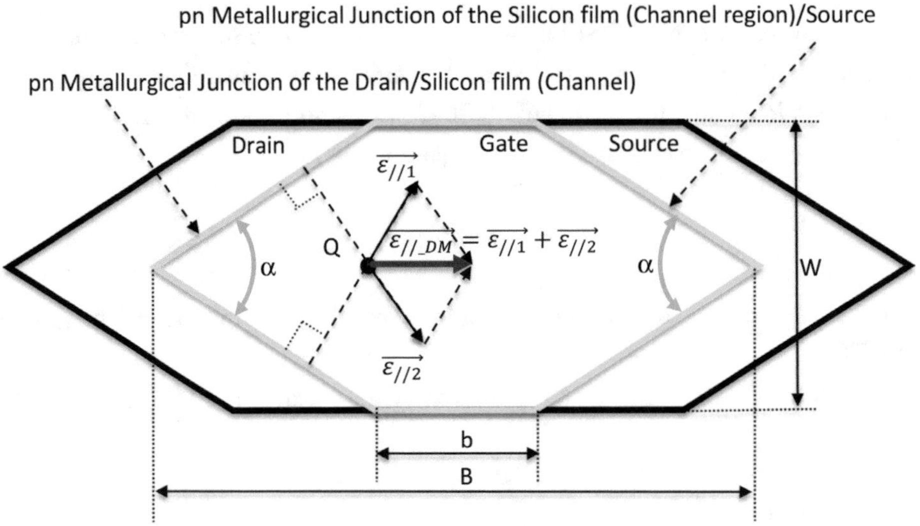

Figure 3.1: Simplified diamond (hexagonal) layout style for MOSFET, where the LEF vectors components and the resultant LEF vector in the point Q are indicated, caused by the V_{DS} bias.

Note that the Diamond MOSFET presents different channel lengths (L_{DM}) along the channel width (W) and varies from b to B, where b $\leq$ LDM $\leq$ B). So, regarding a specific CMOS ICs technology, its effective channel length (L_{eff_DM}) tends to be higher than the L_{min}, motivating its use mainly for analog instead of digital CMOS ICs applications, in which the MOSFETs channel lengths usually tend to be higher than L_{min} [58].

Besides, the Diamond MOSFET can be electrically represented by the parallel connection of many infinitesimal standard MOSFETs with different channel lengths (b $\leq$ L_{DM} $\leq$ B), as shown in Figure 3.2 [30–40].

In Figure 3.2, L_1, L_2, ..., L_i (where i is an integer number, tending to infinite) are the different L of the many infinitesimal rectangular MOSFETs connected in parallel, which compose the Diamond MOSFET structure.

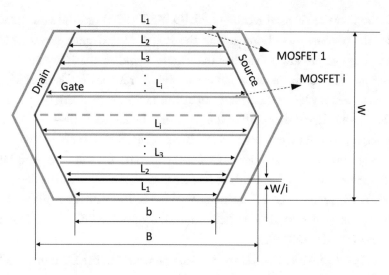

Figure 3.2: The Diamond MOSFET represented by several standard infinitesimal MOSFETs connected in parallel.

Therefore, its equivalent electrical circuit can be illustrated according to Figure 3.3, taking into account that it is a symmetric structure [30–40].

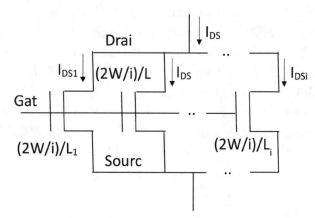

Figure 3.3: The equivalent electrical circuit of the Diamond MOSFET.

In Figure 3.3, I_{DS1}, I_{DS2}, …, I_{DSi} are the different drain currents of each infinitesimal standard rectangular MOSFETs and I_{DS} is the Diamond MOSFET.

Based on Figure 3.3, the effective channel length of the Diamond MOSFET (L_{eff_DM}) is given by Equation (3.3) [30–40].

$$L_{eff\_DM} = \frac{B - b}{ln\left(\frac{B}{b}\right)}$$

(3.3)

Considering that the classical rectangular MOSFET (RM) has the same gate area (A_G) of the DM, where the necessary condition is that the RM L (L_{RM}) must be equal to (b+B)/2, we observe that the L_{eff_DM} are always smaller than those found in the RM counterparts [30–40]. Therefore, the hexagonal gate shape (diamond layout style) for MOSFET is capable of reducing the MOSFET effective channel length (L_{eff}) in comparison to the channel length (L) of a typical rectangular MOSFET, considering that both of them present the same A_G [30–40]. As consequence of this innovative feature, the drain current of the Diamond MOSFET is always higher than the one observed in the conventional rectangular MOSFET, regarding the same A_G and bias conditions. This is justified because the DM I_{DS} tends to further flow in the edges of the device (those infinitesimal MOSFETs with smaller L), according to Figure 3.3. This new effect in the MOSFET with hexagonal gate geometry is named "**PA**rallel connection of **M**OSFETs with **D**ifferent Channel Lengths (**L**) **E**ffect (PAMDLE)" [30–40].

So, the LCE and PAMDLE effects occur simultaneously in the Diamond MOSFET structure and both of them are responsible for remarkably boosting its electrical performance, without adding any extra cost to the current and sophisticated planar CMOS ICs manufacturing processes, which is only a simple layout change [30–40].

As LCE and PAMDLE effects are also capable of improving the transconductance (gm), the gm/I_{DS} ratio and Early voltage (V_{EA}), an experimental study was performed in order to evaluate the frequency response of the common source amplifier (CSA), implemented with hexagonal and standard gate geometries for MOSFETs, by using two different CMOS ICs manufacturing process technologies (350 nm Bulk and 1 μm SOI). The main results of this paper reported that the DM voltage gain (A_{V0}) gains in dB in relation to those observed in the RM counterparts were from 5% to 40% for the Bulk CMOS technology and from 10% to 75% for the SOI CMOS technology by using hexagonal gate shape with α angles ranging from 143.1° to 36.9°, respectively. Furthermore, unit voltage gain frequency (f_T) gains from 7% to 76% for the Bulk CMOS ICs technology and from 20% to 112% for the SOI CMOS technology could be improved by using Diamond MOSFETs with α angles ranging from 143.1° to 36.9°, respectively [34].

A simple analytical model of the Diamond MOSFET drain current (I_{DS_DM}) is given by Equation (3.4), taking into account the LCE and PAMDLE effects and using the IDS of the classical rectangular MOSFET (IDS_RM) as reference, considering that they present the same A_G [34]

$$I_{DS\_DM} = G_{LCE} \cdot GP_{AMDLE} \cdot I_{DS\_RM} \tag{3.4},$$

where the G_{LCE} (LCE Gain) is equal to $\sqrt{2.(1 + cos\alpha)}$ for $0° < \alpha \leq 90°$ and $\sqrt{2 + cos\alpha}$ for $90° \leq \alpha < 180o$ [28, 29], respectively, and G_{PAMDLE} (PAMDLE Gain) is equal to L/L_{eff_DM}, which is valid for all transistor operation regions (subthreshold, triode, and saturation), regarding that they have the same gate areas and overdrive gate voltage ($V_{GT}=V_{GS}-V_{TH}$) [34]. This simple I_{DS_DM} analytical model was validated through experimental data and presented a maximum error smaller than 10%, regarding 100 measurements and 4 pairs of SOI MOSFETs counterparts with the same gate area (DM and CM implemented with 1 μm SOI CMOS ICs technology) [34]. Besides, the I_{DS_DM} gains obtained through experimental data found in relation to the I_{DS_RM} varied from 25% (α = 140°) to 200% (α = 40°), i.e., 3 times higher. These are remarkable results, considering only the gate layout change [34]. Additionally, with this analytical model for Diamond MOSFET, it was possible to distinct the LCE and PAMDLE effects. In this case, the LCE contribution was approximately 27% higher than that found by the PAMDLE. Therefore, regarding those devices' dimensions, the I_{DS_DM} tends to flow in the center farther than in the edges of the channel [34].

This I_{DS_DM} analytical model was also validated through three-dimensional (3D) numerical simulations and considering seven pairs of DM and CM counterparts (same A_G), which were implemented with a 150 nm Fully-Depleted (FD) SOI CMOS ICs technology node [34]. A similar maximum error of 10% was found and therefore this simple I_{DS_DM} analytical model is adequate, ranging from long channel MOSFETs to a 150-nm technology node [34]. Furthermore, although the LCE and PAMDLE are kept active, the I_{DS_DM} gains reduce when the CMOS ICs technology nodes decrease. This happens because the reduction of the PAMDLE effect occurs as a consequence of the L_{eff_DM} not being so reduced in relation to the CM L, as we could observe for the micrometer dimension of the FD SOI MOSFETs [34]. However, the LCE contribution practically has not changed in relation to the higher devices dimensions [34].

Another important investigation was performed between this innovative layout style and a typical one in a high temperature environment (300 K–573 K), regarding the 1-μm SOI CMOS technology node from UCL, Belgium. The main findings of this study reported the hexagonal gate shape for SOI MOSFETs is able to keep the LCE and PAMDLE effects active in high temperature conditions and consequently continue presenting a better electrical performance than the one found in the conventional SOI MOSFET counterpart. Therefore, remarkable improvements of the analog and digital electrical performance of SOI MOSFETs are found, mainly regarding the saturation drain current and unit voltage gain frequency, practically without degrading the voltage gain. So, the diamond layout style for implementing SOI MOSFETs, mainly considering α angles equal to 90°, can be considered an alternative transistor to boost the response frequency of the radio frequency CMOS ICs applications and also the processing velocity of the digital CMOS ICs applications, without adding any extra costs to the current planar SOI CMOS manufacturing processes [35, 36].

An experimental study was also performed, for the first time, regarding Planar Power MOSFETs (PPMs) implemented with diamond layout style (DPPM) with different α angles, as a basic cell, in comparison to the classical rectangular Multifinger PPM (MPPM), regarding the same gate area and bias conditions (Figure 3.4) [37].

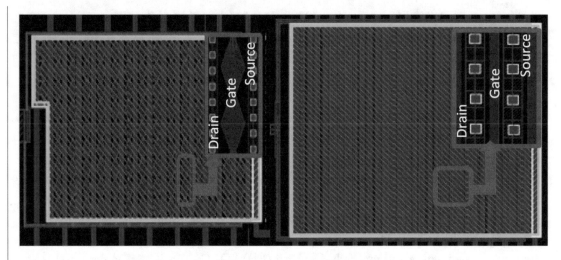

Figure 3.4: Examples of DPPM and MPPM layouts, respectively, with the same gate area.

As a main result of this study, we could observe that the DPPM used as output current driver (switch) in digital CMOS ICs applications is able to remarkably boost the PPMs electrical performance in comparison to the MPPM, considering the same gate area and bias conditions. For instance, we verified that all DPPMs present an $I_{ON}/(W/L)$ 310%, 389% and 713% higher for α angles equal to 135°, 90°, and 45°, respectively, than those found in the MPPM counterparts. Besides, we could note that the DPPM for all α angles presented smaller R_{ON} than those found in the MPPM counterparts, i.e., -55%, -66%, and -80% for the DPPMs with α angles of 135°, 90°, and 45°, respectively, and consequently, the DPPM can enhance the electrical performance in terms of velocity of the digital CMOS ICs and improve the smart power DC/DC converters' efficiency. This can be justified due to the LCE and PAMDLE effects that exist in the diamond MOSFETs [37].

Another study was performed considering the 3D numerical simulations with the diamond layout style, which was applied in FinFETs and compared to the classical rectangular FinFETs, according to Figure 3.5.

This was a conceptual research because today we do not have enough technology to manufacture vertical non-standard rectangular shapes for the MOSFETs gates (FinFETs and MuGFETs) [38]. The objective of this study was to demonstrate how this innovative layout style can also further boost the electrical performance of FinFETs in relation to the classical rectangular FinFETs [38]. The main results of this study reported that the Diamond FinFET is capable of increasing the drain current and the transconductance by approximately 15%, 40%, and 120%, respectively in triode region, regarding the α angles of 136.4°, 90°, and 45.2°, respectively [38]. Besides, in the saturation region, the FinFET I_{DS} is 41%, 139%, and 455% higher for α angles of 136.4°, 90°, and 45.2°, respectively, than those obtained in the classical rectangular FinFETs, considering the same gate area (A_G), aspect ratio (W/L), and bias conditions [38]. This effect is more forceful in Diamond FinFET with an α angle of 45.2° due to the higher LCE in devices with smaller α angles. Therefore, the Diamond FinFET can be considered an alternative device for outstandingly enhancing the current driver of the CMOS ICs applications [38].

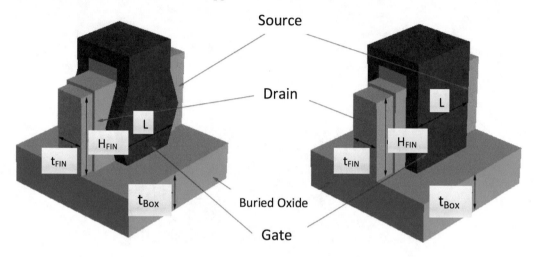

Figure 3.5: Example of a Diamond FinFET and its corresponding classical rectangular FinFET (same gate area and channel length).

Besides these two effects (LCE and PAMDLE), which are capable of boosting the DM electrical performance, an interesting feature of this novel gate structure for MOSFETs is that their resultant LEF lines are curves along the channel length (they are perpendicular to the pn metallurgical junctions between the drain and channel regions, and channel and source regions), according to Figure 3.6, which also illustrates the electric potential (in colors) along the channel length.

It is relevant to highlight that in the specific DM edges between the drain and source regions, named Bird's beak regions (BBRs), the resultant LEF lines are also curves along the channel length (b dimension) [39–40]. Thus, the diamond layout style is capable of electrically deactivating the parasitic MOSFETs of the BBRs and therefore it tends to be more ionizing radiation tolerant (total ionizing dose, TID). This novel effect was named the DEPAMBBRE effect [40].

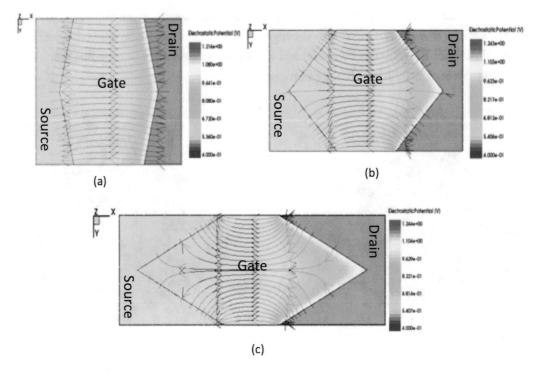

Figure 3.6: The resultant LEF and electric potential (in color) along the channel length of three different FD SOI DMs with the same W (300 nm) and b (150 nm), but different B: 210 nm (a), 390 nm (b) e 630 nm (c), obtained through 3D numerical simulations.

The BBRs are intrinsic in the MOSFETs structures due to the junctions between the thin gate oxide and the isolation oxide, mainly in the Local Oxidation of Silicon (LOCOS) process, despite also existing in a smaller proportion in the Shallow Trench Isolation (STI) process, as illustrated in Figure 3.7 [59, 60].

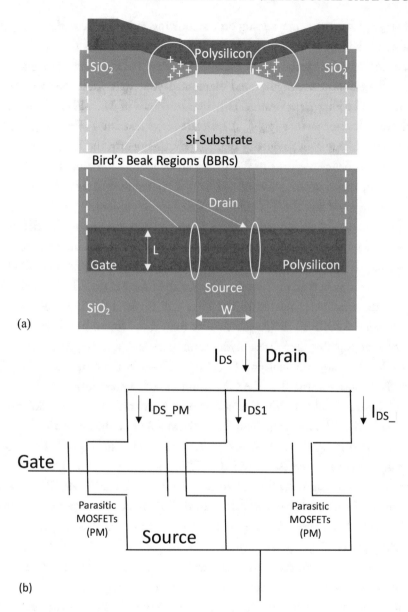

Figure 3.7: The bird's beak regions (BBRs) in the classical MOSFET structure (a) and its corresponding equivalent electrical circuit (b).

When a MOSFET is submitted to an ionizing radiation environment, positive charges are usually induced in the BBRs and therefore the threshold voltages (V_{TH}) of the parasitic MOSFETs associated with these regions reduce. For this reason, these parasitic MOSFETs start to work for V_{GS} values smaller than their original V_{TH} of the main MOSFET and consequently, its leakage drain current (I_{LEAK}) increases. Depending on the ionizing radiation dose, the I_{LEAK} can damage the functionality of the analog or digital CMOS ICs applications [59, 60].

The first comparative study of the protons ionizing radiation effects regarding the TID effects between the classical rectangular and diamond layout styles described that the hexagonal gate shape is capable of improving protons radiation tolerance of MOSFETs remarkably in terms of I_{DS}, maximum transconductance (g_{m_max}), on-state series resistance (R_{DS_ON}), intrinsic voltage gain (A_V), and unit voltage gain frequency (f_T) [32]. Subsequently, the DM improves the current drive, velocity, and frequency response of the analog and digital ICs, mainly for α angles smaller or equal to 90°, in comparison to the conventional MOSFET counterparts [32].

Another study regarding the TID effects caused by X-ray ionizing radiation in the MOS-FETs analog parameters was performed with the Diamond and their respective MOSFETs equivalents (rectangular gate shape). The 350 nm Bulk CMOS ICs manufacturing process of the ON Semiconductor, via MOSIS Educational Program (MEP), was used in this study. One of the main results of this manuscript is that the diamond layout style for MOSFETs is capable of reducing the ionizing radiation influence related to the V_{TH} due to the DEPAMBBRE effect, and also because the Diamond MOSFET BBRs are smaller than those found in the standard rectangular counterparts, considering that they present the same gate area and bias conditions. Besides, we noted that all DMs on-state drain currents (I_{ON}) normalized by the aspect ratio [$I_{ON}/(W/L)$] are always higher than those measured in the CM counterparts, before and after the ionizing radiations; therefore, the LCE and PAMDLE effects remain active after the X-ray radiation procedure. Furthermore, the DM, with an α angle equal to 127°, is able to reduce the off-state drain current (I_{OFF}) normalized by the aspect ratio [$I_{OFF}/(W/L)$] mainly due to the DEMPAMBBRE effect, which electrically deactivates the parasitic MOSFETs in the BBRs. The DM (α = 127°) is not affected by the X-ray ionizing radiations of 76 Mrad due to the DEPAMBBRE effect. Based on this investigation, the hexagonal layout style for MOSFET can be considered a hardness-by-design technique to be applied in current drives and buffers of space CMOS ICs applications [39].

Another experimental comparative study of the Single Event effect (SEE) was performed by using a heavy-ion beam through the 8 UD Pelletron accelerator, between two different MOSFET structures (typical rectangular and diamond layout styles), regarding the 350 nm CMOS ICs technology. For this work, we noted that the SEE has occurred only in the standard rectangular MOSFET. No SEE was observed in the DM. This can be explained mainly due to the DEPAMBBRE effect in the hexagonal MOSFET [39].

PROBLEMS AND QUESTIONS

1. What was the known effect used to boost the longitudinal electric field (LEF) along the channel length of the MOSFETs?

2. Why does the drain current (I_{DS}) and the electrical performance of the MOSFETs increase, when their LEF along of the channel length is boosted? Give mathematical justifications.

3. What is the longitudinal corner effect (LCE)?

4. Explain the Parallel connection between MOSFETs with Different Channel Lengths (L) Effect (PAMDLE).

5. What is the analytical mathematical model of the Diamond MOSFET which takes into account the LCE and PAMDLE effects?

6. What happens with the LCE and PAMDLE effects in MOSFETs when we reduce the CMOS ICs technology node from 1 µm to 150 nm?

7. Is it possible to boost the frequency response of common source amplifiers (CSA) when we use diamond layout style for MOSFET instead of the classical rectangular one? Justify your answer.

8. Are the LCE and PAMDLE effects kept active in a MOSFET structure under high temperature conditions?

9. What would the gains of the diamond layout style be if we could apply it in FinFETs in relation to the classical rectangular layout style?

10. What is the DEPAMBBRE?

11. Are the Diamond MOSFETs recommended to be used in space applications? Give some examples regarding the protons and x-rays ionizing radiations.

CHAPTER 4

Octo Layout Style (Octagonal Gate Shape) for MOSFET

The octagonal gate geometry to implement MOSFETs (Octo-MOSFET, OM) is an evolution of the diamond layout style. It was carefully designed in order to reduce the obtuse corners of the hexagonal gate shape to increase the breakdown voltage (BVDS) and the electrostatic discharge (ESD) tolerance. This can be made by cutting parts of the hexagonal gate geometry by a factor c, according to Figure 4.1 [41–49].

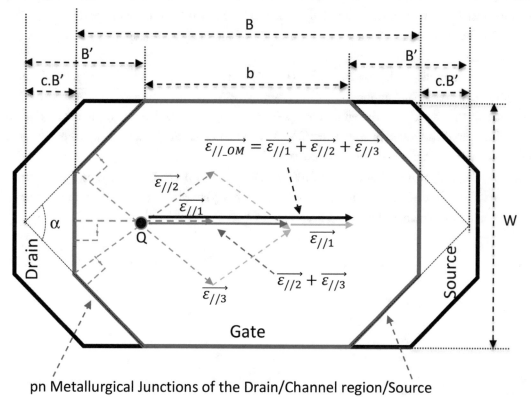

Figure 4.1: Simplified octagonal layout style for MOSFET, indicating the three LEF vectorial components of the same magnitude and the resultant LEF vector at the point Q, caused by the V_{DS} bias.

In Figure 4.1, b (usually defined as L_{min}) and B are the smallest and largest channel lengths of the OM, respectively. B' is the height of the triangular part of the hexagonal gate shape, c is the cutting factor, W is the MOSFET channel width, $\overrightarrow{\varepsilon_{//1}}$, $\overrightarrow{\varepsilon_{//2}}$, and $\overrightarrow{\varepsilon_{//3}}$ are the LEF vectorial components (with the same magnitude) and the corresponding resultant LEF vector ($\overrightarrow{\varepsilon_{//\_OM}} = \overrightarrow{\varepsilon_{//1}} + \overrightarrow{\varepsilon_{//2}} + \overrightarrow{\varepsilon_{//3}}$), due to the V_{DS} [41–49].

This novel layout style for MOSFETs also has the Longitudinal Corner Effect (LCE) incorporated in its structure, which is able to boost its resultant longitudinal electric field ($\overrightarrow{\varepsilon_{//}}$), subsequently the drift velocity of the mobile charge carriers ($\overrightarrow{\vartheta_{//}}$), its drain current, several electrical parameters, and analog and digital figures of merit [41–49].

In this case, there are three LEF vectorial components, which are perpendicular to the PN metallurgical junctions defined between the drain/channel and channel/source regions. Therefore, the resultant LEF ($\overrightarrow{\varepsilon_{//\_OM}}$) along the OM L is higher than the one found in the typical rectangular MOSFET counterpart, regarding the same gate area and bias conditions. Thus, the OM electrical performance is better than the one found in the rectangular MOSFET counterpart.

The resultant LEF depends on the α angle of the octagonal gate shape, as illustrated in Figure 4.1 [41–49].

The magnitude of the resultant LEF ($\overrightarrow{\varepsilon_{//\_OM}}$) can be obtained through the vectorial sum of the three LEF vectorial components and is given by $\overrightarrow{\varepsilon_{//1}} + \sqrt{\varepsilon_{//2}{}^2 + \varepsilon_{//3}{}^2 + 2.\varepsilon_{//2}.\varepsilon_{//3}.\cos(\alpha)}$, which depends on the $\cos(\alpha)$, and therefore, the lower the α angle, the higher the resultant LEF of the OM. The α angle is limited by the planar CMOS ICs manufacturing processes, i.e., there is a minimum α angle for each manufacturing process [41–49].

Analogously to the Diamond MOSFET (DM), the OM can be electrically represented as the parallel connections of infinitesimal standard rectangular MOSFETs with different channel lengths along the channel width (W), according to Figure 4.2 [49].

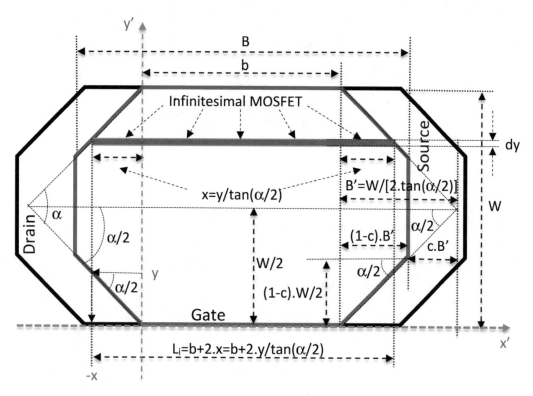

Figure 4.2: Electrical representation of an OM considering many infinitesimal MOSFETs connected in parallel, indicating all related dimensions in order to model its effective channel length.

In Figure 4.2, y' and x' are the Cartesians axes, x and y are dimensions used to model the channel length of the infinitesimal MOSFET, and L_i and dy are respectively the channel length and width of the infinitesimal MOSFET. Besides, note that when $0 \leq y \leq (1-c).(W/2)$, L_i varies $[Li=b+2.x= b+2.y/\tan(\alpha/2)]$, and L_i is constant and equal to B for $(1-c).(W/2) \leq y \leq W/2$, according to Figure 4.3 [49].

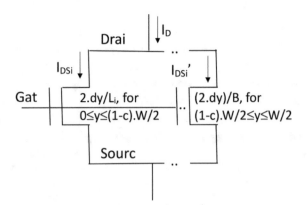

Figure 4.3: Equivalent electrical circuit of the Octo MOSFET.

In Figure 4.3, I_{DSi} and I_{DSi}' are the drain currents of the infinitesimal MOSFETs connected in parallel to the OM structure, and I_{DS} is the OM drain current.

Based on Figure 4.3, the aspect ratio of the Octo MOSFET (W/L_{eff_OM}, where L_{eff_OM} is its effective channel length) can be obtained via Equation (4.1) [49].

$$
\frac{W}{L_{eff\_OM}} = 2. \int_{0}^{(1-c).\frac{W}{2}} \frac{dy}{L_i} + 2. \int_{(1-c).\frac{W}{2}}^{\frac{W}{2}} \frac{dy}{B} =
$$

$$
= 2. \int_{0}^{(1-c).\frac{W}{2}} \frac{dy}{b + \dfrac{y}{tan\left(\frac{\alpha}{2}\right)}} + 2. \int_{(1-c).\frac{w}{2}}^{\frac{W}{2}} \frac{dy}{b + \dfrac{(1-c).W}{tan\left(\frac{\alpha}{2}\right)}} = \tag{4.1}
$$

$$
= tan\left(\frac{\alpha}{2}\right).\ln\left(1 + \frac{W}{b.tan\left(\frac{\alpha}{2}\right)}\right) + \frac{c.W}{b + \dfrac{(1-c).W}{tan\left(\frac{\alpha}{2}\right)}}
$$

So, the OM L_{eff_OM} is given by Equation (4.2) [49].

$$
L_{eff\_OM} = \frac{1}{\dfrac{1-c}{B-b}\ln\left(\dfrac{B}{b}\right) + \dfrac{c}{B}} \tag{4.2}
$$

Note that if c equals 100% (c=1), then L_{eff_OM} is equal to the L of the standard rectangular MOSFET (RM), which is equal to B. And, if c is equal to 0% (c=0), L_{eff_OM} is equal to L_{eff_DM} [Equation (3.3)].

Considering that the RM has the same gate area (A_G) as the OM, where the necessary condition is that the RM L (L_{RM}) must be equal to (b + 2B)/3, regarding Equation (4.2), we note that the L_{eff_OM} are always smaller than those found in the RM counterparts. Therefore, the octagonal layout style for MOSFET is also capable of reducing the MOSFET effective channel length (L_{eff}) in comparison with the L_{RM}, considering that both of them present the same A_G [49]. As a consequence of this particular feature of the OM, its drain current (I_{DS_OM}) is always higher than the one observed in the classical rectangular MOSFET regarding the same A_G and bias conditions [49]. This is justified because the I_{DS_OM} tends to further flow on the edges of the device (the smallest L), according to Figures 4.2 and 4.3 [49]. Therefore, the OM also presents the PAMDLE effect, the same way that the Diamond MOSFET does [49].

So, both the LCE and PAMDLE effects happen concurrently in the OM structure and are responsible for boosting its electrical performance remarkably, without increasing any cost to the current and sophisticated planar CMOS ICs manufacturing processes [49].

Analogously to the DM, a simple analytical model of the OM drain current (I_{DS_OM}) is given by Equation (4.3), taking into account the LCE and PAMDLE effects [49]

$$I_{DS\_OM} = G_{LCE} \cdot G_{PAMDLE} \cdot I_{DS\_RM} \qquad (4.3),$$

where G_{LCE} [$=(\sqrt{2(1+\cos\alpha)} +1)$ for $0 \leq \alpha \leq 90°$ and $(\sqrt{2+\cos\alpha} + 1)$ for $90 \leq \alpha \leq 180°$] and G_{PAMDLE} ($=L/L_{eff\_OM}$) are the OM I_{DS} gains related to the LCE and PAMDLE effects in relation to the RM counterpart (same gate area) in all operation regions (subthreshold, triode, and saturation), regarding the same overdrive gate voltage ($V_{GT}=V_{GS}-V_{TH}$), respectively, and the I_{DS_RM} is the RM I_{DS} [49].

By applying this simple analytical model for the OM, it is possible to differentiate the LCE from the PAMDLE effect, as it was similarly done for the Diamond MOSFETs [49].

The first work related to the OM performed a comparative study by means of three-dimensional (3D) numerical simulations between the Octo, Diamond, and typical rectangular MOSFETs regarding the same bias conditions. In this study, the OM was also considered an alternative device for analog CMOS ICs applications due to the LCE effect, which is capable of improving the MOSFET electrical performance in terms of saturation I_{DS} (I_{DS_sat}), gm, and on-state resistance (R_{ON}) [42]. In this investigation, the PAMDLE was not taken into consideration because it had not been discovered prior to this publication.

After the first experiment, we performed an experimental comparative study between the three different layout styles (Octagonal, Diamond, and rectangular gate geometries), regarding devices manufactured in the WINFAB clean rooms of the Université catholique de Louvain (UCL). The technological parameters of these devices are: gate oxide (t_{ox}), silicon film (t_{Si}), and buried box (t_{BOX}) thicknesses equal to 30 nm, 80 nm, and 390 nm, respectively, and the MOS-FET channel and drain/source doping concentrations are equal to 6.10^{16} cm^{-3} and 1.10^{20} cm^{-3}, respectively. This study reported that the OCTO SOI MOSFET (OSM) with an α angle equal to 53.1° and c equal to 25% practically presents the same $I_{DS}/(W/L)$ (2.2% smaller) than the one observed in the Diamond SOI MOSFET (DSM) with an α angle also equal to 53.1°, but with an OSM gate area (A_G) 27% smaller than the one found in the DSM counterpart, regarding the same bias conditions. Besides, we observed a remarkable 61% OSM (α=126.9° and c=25%) gate area reduction to drive the same Rectangular SOI MOSFET (RSM) $I_{DS}/(W/L)$. Furthermore, the octagonal layout style for implementing SOI MOSFETs was capable of significantly boosting the maximum transconductance (gm_{max}) in 137%, the gm/I_{DS} ratio in 47%, and the unit voltage gain frequency (f_T) in 128% in relation to those found in the classical rectangular SOI MOSFETs (CSMs) counterparts, considering the same gate area (A_G) and bias conditions. Additionally, the OSM voltage gain is practically the same as the one found in the CSM counterpart in strong inversion regime. All these advantages found can be justified on account of the LCE and PAM-DLE effects in the OSM structure. Therefore, based on this first experimental study, we were able to conclude that the octagonal layout is also an alternative technique to improve the integration (die area reduction) and the analog electrical performance, mainly focusing on the radio frequency (RF) of SOI CMOS ICs applications [43].

Another study on the octo layout style confirmed that it can remarkably boost the RF SOI CMOS ICs application performance in terms of frequency response [44]. Besides, the OSM f_T is practically the same as the one found in DSM regarding all inversion regime regions, but with a significant gate die area reduction of 27%. Furthermore, the OSM and DSM f_T were practically 128% higher than the one measured in the CSM counterpart, considering a $I_{DS}/(W/L)$ of 6μA, in strong inversion regime. The higher OSM1 g_m in relation to the CSM homologous justifies this result, considering the same bias conditions. Additionally, the OSM A_G was about 68% smaller in relation to the one measured in the CSM counterpart considering the same f_T. Therefore, we can significantly boost the f_T and reduce the A_G by using the octagonal layout style for MOSFETs in RF SOI CMOS ICs [global positioning system (GPS), wireless local area network (WLAN), bluetooth communication interfaces, and high speed analog-digital converters (ADCs)]. This is achieved by a simple gate layout change in the MOSFETs, without adding any extra cost to the current planar SOI CMOS ICs manufacturing processes [44].

The DEPAMBBRE effect of the Diamond-MOSFET also occurs in the Octo-MOSFET structure, i.e., the resultant LEF lines are curves along the channel length, which are perpendicular to the pn metallurgical junctions between the drain and channel regions and between the channel and source regions, as illustrated in Figure 4.4. It also shows the electric potential (in colors) along channel length [49].

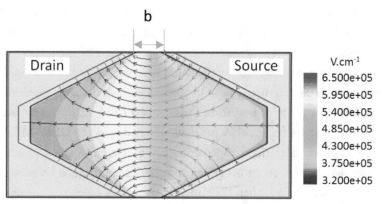

Figure 4.4: The resultant LEF and electric potential (in colors) along the channel length of an Octo-MOSFET, obtained through a 3D numerical simulation.

Figure 4.4 shows that the resultant LEF lines are also curves along the b dimension in the BBRs of the OM [49]. Consequently, the octagonal layout style is also capable of electrically deactivating the parasitic MOSFETs of the BBRs, and therefore it tends to present a higher ionizing radiation robustness than the RM counterpart, regarding the same A_G [49].

Several experiments have been performed with octagonal gate shape for MOSFETs, taking into account the ionizing radiation effects in their main analog and digital parameters and figures of merit. All these studies demonstrate that the Octo-MOSFET is capable of boosting the radiation ionizing tolerance in relation to the RM counterparts, regarding the same bias conditions [45–49].

In the first study in an ionizing radiation environment, the Octo-MOSFET (α=53.1° and c=25%) presented a superior electrical behavior, taking into account the TID effect degradation in terms of threshold voltage (84.9%), subthreshold slope (SS) (-7.7%), and R_{ON} (2.4 times) in relation to the RM homologous, regarding the same A_G and bias conditions. This is explained due to smaller Octo-MOSFET BBRs dimensions in relation to the ones found in the RM counterpart, as well as the DEPAMBBRE effect. Furthermore, we observed that the Octo-MOSFET leakage drain current (I_{LEAK}) decreases 2.69 times in relation to the RM counterpart, after the TID of 600 krad [45]. This is justified because the area of the pn metallurgical junctions between the drain-channel and channel-source regions of the Octo-MOSFET, due to its longer perimeter, are larger than the conventional rectangular SOI MOSFET counterpart. Consequently, the resultant LEF of the Octo-MOSFET is smaller in these pn junctions than in the RM counterpart. As the ionizing radiation effects in MOSFETs depend on the electric field, smaller resultant LEF in these regions lead to smaller ionizing radiation effects in the OSM I_{LEAK}, even though it would tend to increase due to the larger area of pn metallurgical junctions of the Octo MOSFET in relation to its counterpart [47, 49]. Therefore, the ionizing radiation influence due to the smaller resultant LEF seems to be more relevant in relation to the larger area of the pn metallurgical junctions.

The second study confirmed that the octagonal layout style is able to increase the MOSFET electrical performance and improve the X-ray radiation robustness of the analog SOI CMOS ICs applications as it is capable of producing smaller variations, mainly in the threshold voltage (40%), saturation drain current (106%), gm/I_{DS} ratio in the moderate and strong inversion regimes (8%), and voltage gain (10%) than those observed in the standard MOSFET counterpart, regarding the same bias conditions. These results could be justified mainly due to the DEPAMBBRE effect [46].

Another investigation performed an experimental comparative study of the TID effects in SOI MOSFETs implemented with the octagonal and classical rectangular layout styles. The Octagonal SOI MOSFET evidenced a higher TID tolerance, and it is capable of keeping the LCE and PAMDLE effects active after the X-ray radiation exposure, taking the I_{ON}/I_{OFF} ratio into account. It is relevant to highlight that the OSM needs a smaller back-gate bias than the one needed in the CSM counterpart to reestablish the pre-radiation conditions of its parameters (threshold voltage and subthreshold slope) because of the DEPAMBBRE effect [47, 48].

Due to the particular characteristics of the intrinsic Bird's beak regions, responsible for giving rise to the DEPAMBBRE effect, this innovative layout style can remarkably reduce the total ionizing dose (TID) effects, boost the drain current at least twice, and enhance R_{ON}, I_{ON}, I_{OFF}, and I_{ON}/I_{OFF} due to the LCE and PAMDLE effects [48–50].

In conclusion, this innovative octagonal layout style for implementing MOSFETs is a promising hardness-by-design strategy to be applied in digital CMOS ICs applications. It is especially favorable as a switching element in the integrated transceivers and smart-power DC/DC converters operating in ionizing radiation environments. Besides, this layout technique does not add any extra cost to the current and sophisticated planar CMOS ICs manufacturing processes.

PROBLEMS AND QUESTIONS

1. Why was the octagonal gate shape for MOSFETs specially designed?

2. How many LEF vectorial components can we find in the OM? Justify.

3. Analyze the L_{eff_OM} Equation in relation the cut factor (c). Also compare the L_{eff_OM} in comparison to the RM L.

4. What is the analytical mathematical model of the Octagonal MOSFET that takes into account the LCE and PAMDLE effects?

5. Explain the DEPAMBBRE effect in the OM.

6. In your opinion, can the OM be used in space applications? Justify your answer.

7. Regarding the experimental studies already performed with the OSM, describe some electrical performance advantages in relation to the standard rectangular MOSFET counterpart.

8. Why is the OSM I_{LEAK} less influenced by the ionizing radiation in comparison to the standard rectangular MOSFET counterpart?

9. Considering the experimental studies already performed with OSM in ionizing radiation environments, describe some octagonal transistor advantages regarding ionizing radiation tolerance in relation to the standard rectangular MOSFET counterpart.

CHAPTER 5

Ellipsoidal Layout Style for MOSFET

The octagonal layout style for MOSFETs was proposed to further improve the BV_{DS} and ESD tolerance of the Diamond MOSFET. However, the Octo MOSFET continues presenting corners in its structure [41–49], which are smaller than those found in the DMs. It is still possible to further boost the BV_{DS} and the EDS robustness of the Octo MOSFETs, for instance, by using the ellipsoidal gate geometry to implement the MOSFET [50–52]. Figure 5.1 illustrates two examples of Ellipsoidal MOSFETs, which indicates the resultant LEF vectors and their vectorial components, perpendicular to the pn metallurgical junctions between the drain and channel regions and the channel and source regions. In Figure 5.1.a, there are three LEF vectorial components at the point P, which belongs to the line segment defined by the F_1 and F_2 focuses of the ellipsoidal geometry. In addition to this, only two LEF vectorial components are taken into account out of this line segment (Figure 5.1.b).

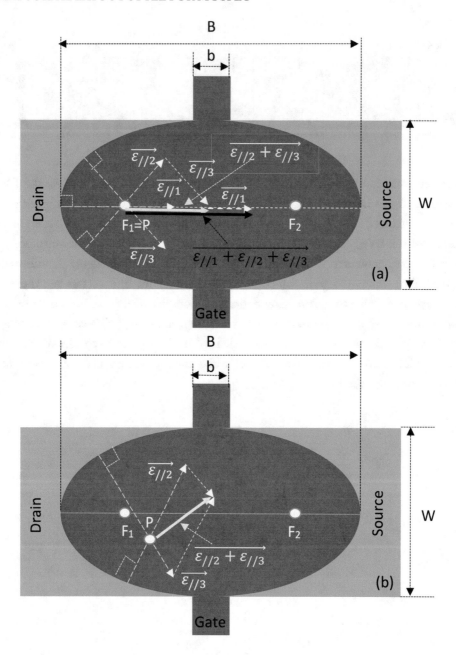

Figure 5.1: Simplified ellipsoidal MOSFET layouts, which indicate the resultant LEF vectors and the LEF vectorial components at any point P of the line segment, defined by the focuses of the ellipse (a), and at a point P out of this line segment, which has only two LEF vectorial components (b), with the same magnitude caused by the V_{DS}.

In Figure 5.1, b (usually defined as L_{min}) and B are respectively the smallest and largest channel lengths of the Ellipsoidal MOSFET, W is the MOSFET channel width, $\overrightarrow{\varepsilon_{//1}}$, $\overrightarrow{\varepsilon_{//2}}$, and $\overrightarrow{\varepsilon_{//3}}$ are the LEF vectorial components with the same magnitude at the point P and the corresponding resultant LEF vectors due to the drain to source voltage (V_{DS}) [50–52].

So, analogously to the DM and OM, the resultant LEF magnitude in the ellipsoidal MOSFET is higher than the one found in the classical rectangular transistor, regarding the same gate area and bias conditions. Thus, the ellipsoidal gate shape is also capable of boosting the mobile carriers' velocity along the channel length and consequently improving the drain current of the MOSFETs and several analog and digital electrical parameters and figures of merit [50–52].

When the author started to investigate the best gate geometric shape for implementing MOSFETs, he found some studies regarding the ellipsoidal shape in engineering. An interesting study was published regarding different gate geometries for target tracking systems in the control system area, taking into account a two-dimensional plane. This study concluded that an ellipsoidal gate computed with Chi-square distribution tables is the optimal shape if compared to the rectangular and circular gates because the probability of the target falling into the ellipsoidal gate is the highest when the volume of the gate is constant [66]. Other research found in nanoelectronics investigated the gate shape effect in the geometric factor (GF) of surrounding gate field-effect transistor (SGFET) with a silicon ellipse shape, taking into account the dispersions of the manufacturing processes. It was verified that for a device with a GF smaller than 1, this shape is promising regarding the analog CMOS ICs due to the short-channel effect being suppressed in these conditions. However, a transistor with a higher GF, such as 2, is more suitable for digital CMOS ICs applications on account of the transient response relying on the MOSFET charge/discharge capability [67]. Another interesting study proposed a model to the Short-Channel Effect (SCE) in the elliptical Gate-All-Around (GAA) MOSFET by using the effective radius (R_{eff}) concept [68]. It is verified that the elliptical GAA MOSFETs that have the same R_{eff} present almost the same SCE and ratio between the on-state and off-state drain currents (I_{on}/I_{off}) [68]. Additionally, another study was published regarding the impact of elliptical cross-section diameter dispersion on the propagation delay of multi-channel GAA MOSFET based on inverters, which proved that even long channel elliptical devices can offer a significant reduction of circuit delay by properly tuning the effective diameter and number of channels [69]. Therefore, besides the investigations performed with the ellipsoidal shapes in different engineering areas, we can further explore this layout style for MOSFETs as it is capable of preserving the LCE, PAMDLE, DEMPAMBBRE, and DRECLE effects of the Diamond and Octo-MOSFETs, which have already been reported in this book.

Analogously to the analyses performed with the Diamond and Octo-MOSFETs, the Ellipsoidal MOSFET (EM) can be divided into infinitesimal standard rectangular MOSFETs with different channel lengths, as illustrated in Figure 5.2 [51, 52].

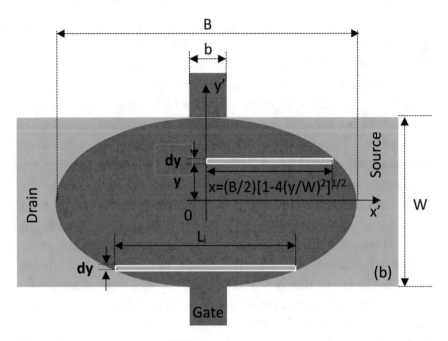

Figure 5.2: Physical representation of an EM considering numerous infinitesimal standard rectangular MOSFETs connected in parallel, indicating all related dimensions in order to model its effective channel length.

In Figure 5.2, y' and x' are the Cartesians axes, x and y are dimensions used to model the channel length of the infinitesimal standard rectangular MOSFETs, and L_i and dy are respectively the channel length and width of the infinitesimal typical rectangular MOSFETs connected in parallel, which compose the EM.

Therefore, the equivalent electrical circuit of the EM is illustrated in Figure 5.3.

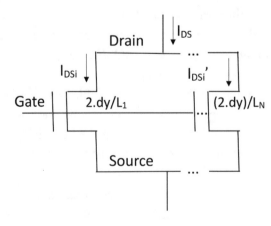

Figure 5.3: Equivalent electrical circuit of the Ellipsoidal MOSFET.

I_{DSi} and I_{DSi}' are the drain currents of the infinitesimal standard rectangular MOSFETs with different channel lengths (L_1, ..., L_N, where N is an integer number that tends to infinite) connected in parallel to the EM structure, and I_{DS} is the EM drain current.

Based on Figure 5.3, the aspect ratio (geometric factor) of the Ellipsoidal MOSFET (W/L_{eff_EM}, where L_{eff_EM} is its effective channel length) can be obtained through Equation (5.1) [51].

$$\frac{W}{L_{eff\_EM}} = 2. \int_0^{y, for\ x=\frac{b}{2}} \frac{dy}{L_i} \tag{5.1}$$

The channel length of an infinitesimal standard rectangular MOSFET that composes the EM structure (L_i) can be obtained via the ellipse equation, as indicated in Equation (5.2) [51].

$$\frac{x^2}{(B/2)^2} + \frac{y^2}{(W/2)^2} = 1 \tag{5.2}$$

So, replacing x with $L_i/2$ in Equation (5.2), L_i is given by Equation (5.3) [51].

$$L_i = B\sqrt{\left(1 - \frac{4.y^2}{W^2}\right)} \tag{5.3}$$

The inferior and superior integration limit ranges from y=0 with x=B/2 to y with x=b/2, where y is calculated by replacing the value b/2 in the x variable via Equation (5.2).

$$y = \frac{W}{2}\sqrt{\left(1 - \frac{b^2}{B^2}\right)} \tag{5.4}$$

Thus, replacing Equations (5.3) and (5.4) in Equation (5.1), we obtain the aspect ratio of the Ellipsoidal MOSFET, as indicated in Equation (5.5).

$$\frac{W}{L_{eff\_EM}} = \frac{2}{B} . \int_0^{\frac{W}{2}.\sqrt{1-\frac{b^2}{B^2}}} \frac{dy}{\sqrt{1 - 4\frac{y^2}{W^2}}} \tag{5.5}$$

In order to solve Equation (5.5), we define $z^2=4(y/W)^2$ and subsequently z=2y/W. Besides, we can calculate dz, which is equal to 2dy/W, and consequently dy=(W/2).dz. Based on this information, the aspect ratio is given by Equation (5.6).

$$\frac{W}{L_{eff\_EM}} = \frac{2}{B} \cdot \int_{0}^{\frac{W}{2} \cdot \sqrt{1-\frac{b^2}{B^2}}} \frac{\frac{W}{2} \cdot dz}{\sqrt{1-z^2}} = \frac{2}{B} \cdot \frac{W}{2} \int_{0}^{\frac{W}{2} \cdot \sqrt{1-\frac{b^2}{B^2}}} \frac{dz}{\sqrt{1-z^2}} \tag{5.6}$$

Solving Equation (5.6), we can obtain Equation (5.7).

$$\frac{W}{L_{eff\_EM}} = \frac{W}{B} \int_{0}^{\frac{W}{2} \cdot \sqrt{1-\frac{b^2}{B^2}}} \frac{dz}{\sqrt{1-z^2}} = \frac{W}{B} \left(\sin^{-1} \frac{z}{1} \right)_{0}^{z} = \frac{W}{B} \sin^{-1} \left(\frac{2 \cdot y}{W} \right)_{0}^{\frac{W}{2} \cdot \sqrt{1-\frac{b^2}{B^2}}} =$$

$$= \frac{W}{B} \cdot \sin^{-1} \left(\frac{2 \cdot \left(\frac{W}{2} \cdot \sqrt{1-\frac{b^2}{B^2}} \right)}{W} \right) = \frac{W}{B} \cdot \sin^{-1} \left(\sqrt{1 - \left(\frac{b}{B} \right)^2} \right) \tag{5.7}$$

Therefore, L_{eff_EM} can be obtained from Equation (5.7), which takes into account the PAM-DLE effect, according to Equation (5.8).

$$L_{eff\_EM} = \frac{B}{\sin^{-1} \left(\sqrt{1 - \frac{b^2}{B^2}} \right)} \quad , \tag{5.8}$$

In order for a standard rectangular MOSFET to present the same gate area of an EM, given by $[(\pi.W.B)/4]$, the necessary condition is that L be equal to $(\pi.B)/4$. We can observe that the L_{eff_EM} are always smaller than those found in the standard rectangular MOSFET counterparts if we compare two MOSFETs with the same gate areas and channel widths, where one of them is an EM and the other is a typical rectangular MOSFET, according to Table 5.1. This occurs due to the PAMDLE effect.

Table 5.1: Different CM and EM dimensions regarding the same gate area and the Leff_EM reductions in relation to the CMs Ls

CM			EM				
W	L	A_G	W	b	B	L_{eff_EM}	$(L_{eff\_EM} - L)/L$
[µm]	[µm]	[µm²]	[µm]	[µm]	[µm]	[µm]	[%]
5.95	20.00	119.0	5.95	1.05	25.465	16.65	-16.8
5.95	15.00	89.25	5.95	1.05	19.099	12.60	-16.0
5.95	9.975	59.35	5.95	1.05	12.701	8.53	-14.4
5.95	7.00	41.65	5.95	1.05	8.13	6.14	-12.2
5.95	4.673	27.81	5.95	1.05	5.95	4.27	-8.6
5.95	4.025	23.95	5.95	1.05	5.125	3.76	-6.5
5.95	3.142	18.69	5.95	1.05	4.001	3.07	-2.4

The ellipsoidal layout style is capable of reducing the MOSFETs' effective channel length (L_{eff_EM}) in comparison with the channel length of a typical rectangular MOSFET (L_{RM}), considering that both of them present the same A_G [51, 52]. Therefore, the EM drain current (I_{DS_EM}) is always higher than the one observed in the standard rectangular MOSFET regarding the same A_G and bias conditions. This is justified because the I_{DS_EM} tends to further flow on the edges of the device (where we can find the smallest L), according to Figures 5.2 and 5.3. In conclusion, the EM, the Diamond, and Octo-MOSFETs present the PAMDLE effect [51, 52].

Thus, the LCE and PAMDLE effects occur simultaneously in the EM structure and are capable of remarkably boosting its electrical performance, without adding any additional cost to the current and sophisticated planar CMOS ICs manufacturing processes.

Analogously to the DM and OM, a simple analytical model of the EM drain current (I_{DS_EM}) [51, 52] is given by Equation (5.9), taking into account the LCE and PAMDLE effects

$$I_{DS\_EM} = G_{LCE} \cdot G_{PAMDLE} \cdot I_{DS\_RM} \qquad (5.9),$$

where G_{LCE} is the I_{DS_EM} gain due to the LCE effect. It is approximately equal to $\sqrt{\varepsilon_{//1}^2 + \varepsilon_{//2}^2 + 2.\varepsilon_{//1}^2.\varepsilon_{//2}2.\cos(\alpha)}$, neglecting the longitudinal electric field vectorial component, which corresponds to the line segment defined by the ellipse focuses (F_1 and F_2). G_{PAMDLE} ($=L/L_{eff\_EM}$) is the EM I_{DS_EM} gain related to the PAMDLE effect in relation to the RM counterpart, with the same gate areas, in all operation regions (subthreshold, triode, and saturation), regarding the same overdrive gate voltages ($V_{GT}=V_{GS}-V_{TH}$) [51, 52], respectively, and the I_{DS_RM} is the RM I_{DS} [51, 52].

By applying this simple analytical model for the EM, it is possible to separate the LCE and PAMDLE effects, as it was performed with the Diamond and Octo-MOSFETs.

Analogously to the Diamond and Octo-MOSFETs, the Ellipsoidal MOSFET is also capable of generating the DEPAMBBRE effect in its structure. Consequently, its ionizing radiation tolerance tends to be higher than the one found in the typical rectangular MOSFET counterpart, regarding the same gate area and bias conditions.

As the areas of the pn metallurgical junctions between the drain and channel regions and between the channel and source regions of the EM are also larger than those found in the classical rectangular MOSFET counterpart, regarding the same gate area, the longitudinal electrical field in these regions tends to be smaller than those observed in the RM. Therefore, the EM presents the DLECRE effect and it tends to present a smaller drain leakage current (I_{LEAK}) in comparison to the RM counterpart, considering the same gate area and bias conditions, as it had previously occurred with the Octo-MOSFET, which was reported in Chapter 4. This phenomenon will be investigated in future studies.

Two studies have been performed with Ellipsoidal MOSFET so far. The first study reported the impact of using the ellipsoidal gate geometry in relation to the typical rectangular MOSFET counterpart intentionally in planar CMOS ICs technology [51]. It was based on experimental results from manufactured devices in the 0.35 μm Bulk CMOS technology. This investigation experimentally proved that the ellipsoidal layout style is capable of boosting about 2 and 3.2 times the on-state drain current and the saturation drain current, respectively. The delay time constant in comparison to the typical rectangular MOSFET equivalent was also reduced in about 61%, regarding the same gate area, without adding any additional cost to this current planar CMOS manufacturing process, thanks to the LCE and PAMDLE effects. Furthermore, this study investigated that its f_T is much higher than the one found in the RM equivalent. Thus, the ellipsoidal layout style can be considered an alternative technique (hardness-by-design) to be used in RF CMOS ICs applications, extremely important to congregate the current demand associated with efficiency and cost reduction.

The second study through 3D numerical simulations investigates the use of ellipsoidal layout style regarding the electrical performance of a MOSFET switch. This specific layout style adds two new effects to the MOSFET structure named Longitudinal Corner Effect (LCE) and Parallel Connection of MOSFET with Different Channel Lengths Effect (PAMDLE), which are able to boost the main digital figures of merit in relation to the RM counterpart. For instance, the EM I_{DS} in Triode and Saturation regions are respectively 128% (V_{GT}=1.5 V) and 207% (V_{DS}=1V) higher than those found in the RM counterpart. The main findings of this work demonstrate that the ellipsoidal gate geometry can be considered an alternative layout technique to be used in MOSFET switches to remarkably enhance its electrical performance and, consequently, the performance of the DC/DC converters [52].

Based on all these studies which have been performed so far regarding these innovative layout techniques, the author considers the ellipsoidal layout style to be the best geometric shape for planar MOSFETs, where the minimal channel length is not required by the design, mainly for analog planar CMOS ICs applications. This can be easily explained because this pioneering layout style presents at least four new effects (<u>LCE</u>: it boosts the LCE along channel length; <u>PAMDLE</u>: it reduces the effective channel length of MOSFETs; <u>DEPAMBBRE</u>: it electrically deactivates the parasitic MOSFETs in BBRs; <u>DLECRE</u>: it reduces the drain leakage current of MOSFETs; <u>this device does not present corners</u>: it increases the breakdown voltage and improves the ESD tolerance). Furthermore, this transistor reduces the die area of analog CMOS ICs regarding the same drain current, and it is capable of boosting the electrical performance of MOSFETs without adding any costs to the current and sophisticated planar CMOS ICs manufacturing processes, only layout change.

At this moment, it is essential to point out that the hexagonal, octagonal, and ellipsoidal layout styles were specially designed for implementing MOSFETs to be used in analog CMOS ICs applications because they are not capable of producing channel lengths equal to the minimum dimension (L_{min}) allowed by the manufacturing processes.

The following procedure can be used to transform a standard rectangular MOSFET with a specific aspect ratio (W/L, where $L>L_{min}$) into another with non-typical gate geometric shape (diamond, octagonal, and ellipsoidal):

1. Match the non-standard MOSFET drain current to the one of the typical rectangular gate MOSFETs;

2. Determine the aspect ratio of the non-standard MOSFET regarding the same bias conditions. The aspect ratio value found in the MOSFET with non-typical gate shape is obviously smaller than the one found in the classic rectangular MOSFETs due to the LCE and PAMDLE effects.

PROBLEMS AND QUESTIONS

1. Why were the Ellipsoidal MOSFETs proposed?

2. How many LEF vectorial components are present in the EM? Justify.

3. Develop the L_{eff_EM} Equation.

4. What is the necessary condition for a standard rectangular MOSFET to present the same gate area of an Ellipsoidal MOSFET?

5. What is the analytical mathematical model of the EM that takes into account the LCE and PAMDLE effects?

6. Consider a standard rectangular with an aspect ratio (W/L) of 20 nm, where the channel width and length are respectively 40 nm and 2 nm. Determine the aspect ratio of an Ellipsoidal MOSFET in order to produce the same drain current, considering that the LCE gain (G_{LCE}) and PAMDLE gain (G_{PAMDLE}) are respectively equal to 15% and 10% and both of them present the same bias conditions. Consider that the minimum dimension allowed by the CMOS ICs manufacturing process is equal to 1 nm and both of them present the same threshold voltage (V_{TH}).

7. In your opinion, can the EM be used in space applications? Justify your answer.

8. Considering the experimental studies already performed with the EM, describe some benefits of the ellipsoidal layout style for MOSFETs in terms of electrical performance in comparison to the classical rectangular MOSFET.

CHAPTER 6

Fish Layout Style ("<" Gate Shape) for MOSFET

All layout styles (hexagonal, octagonal, and ellipsoidal) described so far in this book were specially proposed to be used in MOSFETs of the analog CMOS ICs because their effective channel lengths (L_{eff}) are always higher than the minimum dimension allowed (L_{min}) by the CMOS ICs technology node considered.

The fish layout style was specially designed to be applied in MOSFETs to explore the same layout techniques in digital CMOS ICs applications. It presents a gate shape similar to the "less than" mathematical symbol (<) to preserve the LCE effect of the Diamond MOSFET, according to Figure 6.1 [53, 54].

In Figure 6.1, W is the MOSFET channel width, $\overrightarrow{\varepsilon_{//1}}$ and $\overrightarrow{\varepsilon_{//2}}$ are the LEF vectorial components (with the same magnitude) and the corresponding resultant LEF vector ($\overrightarrow{\varepsilon_{//\_T}} = \overrightarrow{\varepsilon_{//1}} + \overrightarrow{\varepsilon_{//2}}$) due to the potential difference between the drain and the source (V_{DS}), α is the angle that defines how obtuse the Fish MOSFET (FM) is, and L_{eff} is its effective channel length [53, 54].

The FM I_{DS} direction is defined by $\overrightarrow{\varepsilon_{//\_T}}$, which is higher than the one found in the RM counterpart, regarding the same gate area and bias conditions. Besides, $\overrightarrow{\varepsilon_{//\_T}}$ is responsible for increasing the drift velocity of the mobile charger carriers along the channel length and consequently the drain current, and also the majority of the analog and digital parameters and figures of merit of the MOSFETs.

The FM can be manufactured with the smallest feature length allowed by the CMOS ICs manufacturing process technology, and it presents an effective channel length (L_{eff_FM}) higher than the one found in the typical rectangular MOSFET counterpart, which is given by $L/\sin(\alpha/2)$, according to Figure 6.1 [53, 54]. Therefore, the fish layout style is capable of increasing its L_{eff}, depending on the α angle value, and also enhancing the figure of merit defined as $I_{DS\_FM}/(W/L_{eff})$. Equation (6.1) shows the ratio between the FM $I_{DS\_FM}/(W/L_{eff})$ and its corresponding classical rectangular MOSFET (RM) counterpart [$I_{DS\_RM}/(W/L)$] [53, 54].

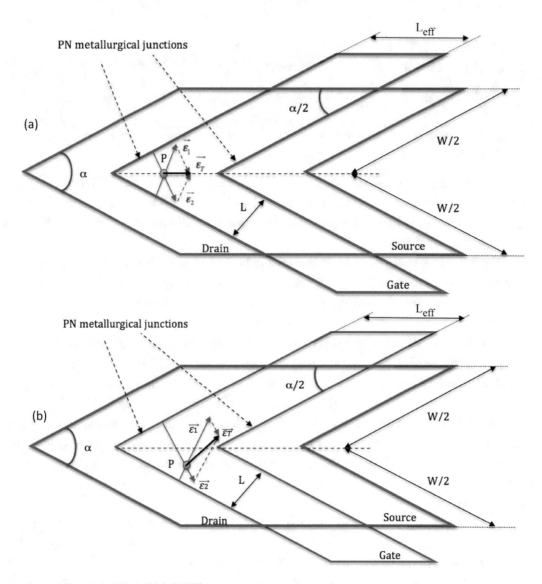

Figure 6.1: Simplified Fish MOSFET layouts, which indicate the resultant LEF vectors and their LEF vectorial components at any point P in the center of the channel (a) and out of this region (b), with the same magnitude produced by the V_{DS}.

$$\frac{\frac{I_{DS\_FM}}{\left(\frac{W}{L_{eff}}\right)}}{\frac{I_{DS\_RM}}{\left(\frac{W}{L}\right)}} = \frac{I_{DS\_FM}}{I_{DS\_RM}}\left[\frac{1}{sin\left(\frac{\alpha}{2}\right)}\right], \quad \text{for } 0 < \alpha < 180^o \quad (6.1)$$

Equation (6.1) shows that FM can generate higher $I_{DS}/(W/L_{eff})$ than the one observed in the RM counterpart. This can be justified due to I_{DS_FM} being higher than I_{DS_RM} on account of the LCE effect, considering that the devices present the same L, W, A_G, and bias conditions. This effect is named "Lengthening of L_{eff} Effect (LECLE)" [53, 54]. This special feature of the fish layout style can be used for current drivers and buffers in digital CMOS ICs applications [53, 54].

Another benefit of this innovative layout style for MOSFET is its reduced sensitivity to the short-channel effect (SCE), since its L_{eff} is larger than the standard rectangular MOSFET counterpart, regarding the same L, W, and A_G [53, 54].

It is relevant to highlight that the magnification of the resultant LEF mainly occurs in the center of the fish gate structure (point P in Figure 6.1) and reduces its level as point P reaches the MOSFET edges. This happens on account of the smaller interaction between the two electrical fields ($\vec{\varepsilon_1}$ and $\vec{\varepsilon_2}$), as illustrated in Figure 6.1 [53, 54].

The LECLE effect that the fish layout style provides to the MOSFETs can be used to increase the integration of digital CMOS ICs applications (logic gates, combinational, sequential integrated circuits, etc.). As an example, the p-type MOSFET aspect ratio $[(W/L)_P]$ of an inverter implemented only with a classical rectangular layout style is calculated by the multiplication of the ratio of the n and p-type mobilities $[(\mu_N/\mu_P)]$ and the n-type MOSFET aspect ratio $[(W/L)_N]$, according to Equation (6.2) [58].

$$\left(\frac{W}{L}\right)_P = \left(\frac{\mu_N}{\mu_P}\right)\left(\frac{W}{L}\right)_N \tag{6.2}$$

Knowing that the n-type MOSFET mobility is at least two or three times higher than the one found in a p-channel MOSFET and adopting the same channel length of both transistors, the channel width (W_p) of the p-type MOSFET is at least two or three times larger than the one found in the n-type transistor. Therefore, the transistors are able to produce the same drain current (I_{DS}) [53, 54].

However, it is possible to decrease the W_p due to the LECLE effect of the fish layout style. This can be achieved by using the classical rectangular layout for the p-channel device and the n-channel Fish MOSFET, according to Equation (6.3) [53, 54]

$$\left(\frac{W}{L}\right)_P = G_{LCE}\left(\frac{\mu_N}{\mu_P}\right)\left(\frac{W}{L_{eff}}\right)_N = G_{LCE}\left(\frac{\mu_N}{\mu_P}\right)\left(\frac{W}{L/\sin(\alpha/2)}\right)_N = G_{LCE}\left(\frac{\mu_N}{\mu_P}\right)\sin\left(\alpha/2\right)\left(\frac{W}{L}\right)_N \tag{6.3},$$

as its simplified analytical model is given by Equation (6.4), which takes into account the LCE and LECLE effects.

$$I_{DS\_DM} = G_{LCE} \cdot \sin\left(\alpha/2\right) \cdot I_{DS\_RM} \tag{6.4}$$

The GLCE (LCE Gain) is approximately equal to $\sqrt{2.(1+\cos\alpha)}$ for $0°<\alpha\leq90°$ and $\overline{\sqrt{2+\cos\alpha}}$ for $90°\leq\alpha<180°$ [28, 29], respectively, and I_{DS_RM} is the I_{DS} of an RM counterpart (with the same gate area) for all transistor operation regions (subthreshold, triode, and saturation), regarding the same overdrive gate voltage ($V_{GT}=V_{GS}-V_{TH}$).

Regarding an inverter gate implemented with a p-type standard rectangular MOSFET and an n-type Fish-MOSFET, Table 6.1 shows that the W_p reductions considering L_{min} and W_n equal to 1 nm, and the ratio between electrons and holes mobilities (μ_n/μ_p) equal to 2.

Table 6.1: Aspect ratio reductions in percentage of the p-type MOSFET when we use an n-type Fish-MOSFET with different α angles to implement an inverter gate

(nm)		(o)		G_{LCE}	(o)		L_{eff_FM} (nm)	(nm)	(nm)	(%)
L_{min}	μ_n/μ_p	α	$2+\cos(\alpha)$	$[2+\cos(\alpha)]^{1/2}$	α/2	sen(α/2)	L/sen(α/2)	W_n	W_p	% WP reduction
1.0	2.0	180	1.00	1.00	90	1.00	1.00	1.00	2.00	0.0
1.0	2.0	175	1.00	1.00	88	1.00	1.00	1.00	2.00	0.1
1.0	2.0	170	1.02	1.01	85	1.00	1.00	1.00	2.01	0.4
1.0	2.0	165	1.03	1.02	83	0.99	1.01	1.00	2.02	0.8
1.0	2.0	160	1.06	1.03	80	0.98	1.02	1.00	2.03	1.4
1.0	2.0	155	1.09	1.05	78	0.98	1.02	1.00	2.04	2.1
1.0	2.0	150	1.13	1.06	75	0.97	1.04	1.00	2.06	2.9
1.0	2.0	145	1.18	1.09	73	0.95	1.05	1.00	2.07	3.6
1.0	2.0	140	1.23	1.11	70	0.94	1.06	1.00	2.09	4.4
1.0	2.0	135	1.29	1.14	68	0.92	1.08	1.00	2.10	5.1
1.0	2.0	130	1.36	1.16	65	0.91	1.10	1.00	2.11	5.6
1.0	2.0	125	1.43	1.19	63	0.89	1.13	1.00	2.12	5.9
1.0	2.0	120	1.50	1.22	60	0.87	1.15	1.00	2.12	6.1
1.0	2.0	115	1.58	1.26	58	0.84	1.19	1.00	2.12	5.9
1.0	2.0	110	1.66	1.29	55	0.82	1.22	1.00	2.11	5.5
1.0	2.0	105	1.74	1.32	53	0.79	1.26	1.00	2.09	4.7
1.0	2.0	100	1.83	1.35	50	0.77	1.31	1.00	2.07	3.5
1.0	2.0	95	1.91	1.38	48	0.74	1.36	1.00	2.04	2.0
L_{min}	μ_n/μ_p	α	$2+\cos(\alpha)$	$[2+\cos(\alpha)]^{1/2}$	α/2	sen(α/2)	L/sen(α/2)	W_n	W_p	% WP reduction
1.0	2.0	90	2.00	1.41	45	0.71	1.41	1.00	2.00	0.0
1.0	2.0	85	2.17	1.47	43	0.68	1.48	1.00	1.99	-0.4
1.0	2.0	80	2.35	1.53	40	0.64	1.56	1.00	1.97	-1.5
1.0	2.0	75	2.52	1.59	38	0.61	1.64	1.00	1.93	-3.4
1.0	2.0	70	2.68	1.64	35	0.57	1.74	1.00	1.88	-6.0

1.0	2.0	65	2.85	1.69	33	0.54	1.86	1.00	1.81	-9.4
1.0	2.0	60	3.00	1.73	30	0.50	2.00	1.00	1.73	-13.4
1.0	2.0	55	3.15	1.77	28	0.46	2.17	1.00	1.64	-18.1
1.0	2.0	50	3.29	1.81	25	0.42	2.37	1.00	1.53	-23.4
1.0	2.0	45	3.41	1.85	23	0.38	2.61	1.00	1.41	-29.3
1.0	2.0	40	3.53	1.88	20	0.34	2.92	1.00	1.29	-35.7
1.0	2.0	35	3.64	1.91	18	0.30	3.33	1.00	1.15	-42.6
1.0	2.0	30	3.73	1.93	15	0.26	3.86	1.00	1.00	-50.0
1.0	2.0	25	3.81	1.95	13	0.22	4.62	1.00	0.85	-57.7
1.0	2.0	20	3.88	1.97	10	0.17	5.76	1.00	0.68	-65.8
1.0	2.0	15	3.93	1.98	8	0.13	7.66	1.00	0.52	-74.1
1.0	2.0	10	3.97	1.99	5	0.09	11.47	1.00	0.35	-82.6
1.0	2.0	5	3.99	2.00	3	0.04	22.93	1.00	0.17	-91.3

Based on Table 6.1, it is possible to remarkably reduce the die area of an p-type MOSFET ranging from 13.4% to 29.3% and also the inverter gate, considering α angles varying from 45° to 60°. This occurs because the contribution of the LCE effect becomes higher than the LECLE effect. Therefore, for the first time, this innovative layout methodology is capable of reducing the die area of digital CMOS ICs applications. This layout technique can be extended for any digital CMOS ICs applications (combinational and sequential ICs). The impact of this new technique use can bring great cost reductions of the digital CMOS ICs (computers, cell phones, embedded electronics, etc.) [53, 54].

Furthermore, considering the analog CMOS ICs applications, where the current mirrors are also calculated based on the geometric factors (aspect ratios) between two or more MOSFETs, we also can apply this same approach in order to reduce their die areas [53, 54].

The references [53] and [54] presented a comparative study between the Fish and the typical rectangular layout styles for MOSFETs via 3D numerical simulations. This work showed that the threshold voltages (V_{TH}), the subthreshold slope, and I_{on}/I_{off} ratio of the MOSFETs were practically the same. It also showed that the fish layout style was able to produce higher $I_{DS}/(W/L_{eff})$, saturation $I_{DS}/(W/L_{eff})$, $gm/(W/L_{eff})$, gm/I_{DS} ratio, voltage gain (A_V) and unit voltage gain frequency (f_T), Early voltage (V_{EA}), and on-state resistance (R_{ON}). Consequently, it can be used as a layout technique for analog and digital CMOS ICs applications, such as amplifiers and current driver/buffer due to the LCE effect, where the Fish MOSFET L_{eff} is larger than the minimum size allowed by the CMOS ICs technology node.

In conclusion, the fish layout style can be considered an alternative layout technique to boost the electrical performance of both analog and digital ICs and mainly reduce the die area and chip cost, as a consequence of the LCE and LECLE effects in the Fish MOSFET structure. Besides, the DEPAMBBRE is also presented in the fish layout style and it can also be used as a hardness-by-design technique to implement digital CMOS ICs to boost the ionizing radiation tolerance of the MOSFETs, focusing on the space applications (future studies).

PROBLEMS AND QUESTIONS

1. Why was the fish gate shape for MOSFETs specially designed?

2. How many LEF vectorial components can we find in the Fish MOSFET? Justify.

3. Analyze the L_{eff_FM} Equation in relation to the α angle. Also compare L_{eff_FM} in comparison to the RM L.

4. What is the analytical mathematical model of the FM that takes into account the LCE and LECLE effects?

5. Regarding the experimental studies already performed with the FM, describe some electrical performance advantages in relation to the typical rectangular MOSFET counterpart.

6. Why would it be interesting to use the fish layout style in digital CMOS ICs applications, taking into account the devices' integration? Justify.

7. Why can the Fish MOSFET be used in space CMOS ICs applications? Justify.

CHAPTER 7

Annular Circular Gate Layout Style for MOSFET

The circular annular layout style makes the MOSFETs an asymmetrical device because the areas of the internal and external regions are different, which are adjacent to the gate region. Figure 7.1 illustrates the Circular Annular MOSFET (CA-M) operating in external drain bias configurations (EDBC), Figure 7.1.a, and internal drain bias configuration (IDBC), Figure 7.1.b, respectively.

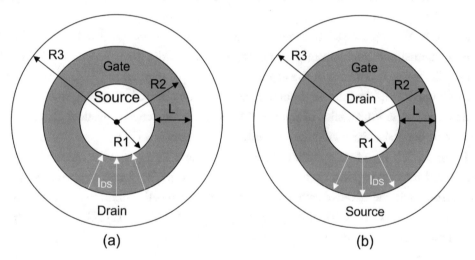

Figure 7.1: Simplified layouts of the CA-M in the external (a) and internal drain bias configurations (b).

In Figure 7.1, I_{DS} is the drain current of the CA-M, R1 is the radius of the circle that defines the beginning of the gate region, R2 is the radius of the circle that defines the end of the gate region, L is the channel length given by R2-R1, and R3 is the radius of the circle that defines the end of the external region of the CA-M, which can be the drain region or the source region, depending on the drain bias configuration (external or internal) [70].

The drain current (I_{DS}) of the CA-M is radial, different from the classical rectangular MOS-FET. Consequently, it can be used to improve the match between devices because in 360° of a gate region, there are always two I_{DS} components in opposite directions, which can compensate the systematic and random errors due to the CMOS ICs manufacturing process [25, 26].

The relation between the geometric factors (aspect ratio) of the standard rectangular MOSFET and the CA-M is given by Equation (7.1) [70].

$$f_g = \left(\frac{W}{L}\right)_{rectangular} = \left[\frac{2.\pi}{ln\left(\frac{R_2}{R_1}\right)}\right]_{Circular\ Anular} \tag{7.1}$$

By analyzing Equation (7.1), the f_g of CA-M is limited at about 15.5 with a small total die area, respecting the design rules, where the allowed dimensions to implement the transistors are usually multiples of the middle of the minimum size allowed by the CMOS ICs technology (lambda, λ). Besides, its total area increases significantly to produce smaller f_g than this value of 15.5, limiting its application.

As the CA-M is geometrically asymmetric, its electrical behavior is also different when it is operating in external and internal drain bias configurations, as the series resistance of the source region is smaller when it operates in the external drain bias configuration (EDBC) than the one found when it operates in the internal drain bias configuration (IDBC), according to Figure 7.2. In this figure, we are presenting a slice of the CA-M and their respective equivalent electric circuits for each type of the drain bias configurations (internal and external) [71].

In Figure 7.2, V_{DD} is the power supplier; R_{S_EXT} and R_{S_INT} are the series resistances of the external and internal regions, respectively; V_{GS_ED}', V_{GS_ID}', V_{DS_ED}', V_{DS_ID}', I_{DS_ED}, and I_{DS_ID} are respectively the effective gate to source voltage, the effective drain to source voltage, and the drain current of the CA-M, when it is operating in the EDBC and IDBC; and V_{GS}, V_{DS}, are the gate to source, and the drain to source voltages, respectively [71].

Observe that the R_{S_INT} of Figure 7.2 is higher than the R_{S_EXT} because they are inversely proportional to their transversal section areas ($A\_INT > A\_EXT$), considering that they have the same length.

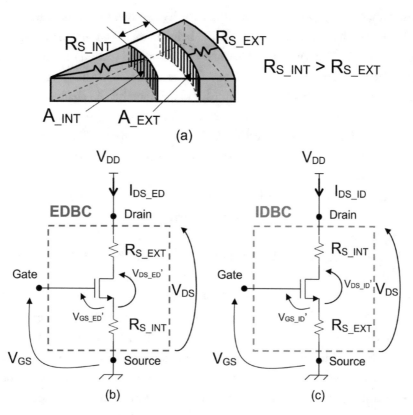

Figure 7.2: A slice of the CA-M (a) and the equivalent electric circuits in the EDBC (b) and in IDBC (c), respectively.

Equations (7.2) and (7.3) represent the effective gate to source voltages due to the influence of the source series resistance of the CA-M, when it is operating in EDBC and IDBC, respectively.

$$V_{GS\_ED'} = V_{GS} - R_{DS\_INT} \cdot I_{DS\_ED} \tag{7.2}$$

$$V_{GS\_ID'} = V_{GS} - R_{DS\_EXT} \cdot I_{DS\_ID} \tag{7.3}$$

Based on Equations (7.2) and (7.3), neglecting the channel length modulation effect in the drain current of the CA-M, the effective gate to source ($V_{GS\_DE'}$) when it is working in the EDBC is smaller than the one observed when it is operating in the IDBC due to the R_{DS_INT} being higher than the R_{DS_EXT}. Subsequently, the drain current of CA-M is higher when it is functioning in the IDBC than the one when it is operating in the EDBC [71].

Additionally, the I_{DS_ID} is also higher than the I_{DS_ED} due to the longitudinal electric field (LEF) in CA-M when it is operating in the IDBC being higher than the one measured, when it is functioning in the EDBC. This is justified because the transversal section area of the internal region (A_{INT}) is smaller than the one from the transversal section area of the external region (A_{ENT}) of the CA-M. As a result of this effect, the impact ionization and the channel length modulation effect (pinch-off displacement) in this transistor when it is working in the IDBC is higher than the one found when it is functioning in the EDBC. Therefore, all these effects are capable of further boosting the drain current of the CA-M when it is operating in the IDBC in relation to it when it is working in the EDBC [22].

We usually use the CA-MOSFET in the IDBC to reduce the drain parasitic capacitance. Consequently, it increases the unit voltage gain frequency (f_T) and reduces the Miller capacitance effect in the amplifiers [25, 26].

Thus, the use of circular annular MOSFETs in the analog and digital CMOS ICs applications must be done in a careful way due to its geometric asymmetry.

In planar power MOSFETs (PPMs), it is possible to use the CA-M as a basic-cell, where the gate regions of several transistors can be connected to each other in order to significantly increase the integration factor, given by ratio of the aspect ratio and the total die area (Overlapping-Circular Gate Transistor, O-CGT) in relation to the multifinger layout style of PPMs [25, 26].

The circular annular layout style for MOSFETs is usually used in order to boost the radiation tolerance because it does not present the Bird's beak regions in its structure [23, 24, 72].

PROBLEMS AND QUESTIONS

1. Why is the circular annular MOSFET an asymmetric device?

2. Can the CA-MOSFET be considered an alternative device to improve the matching between devices? Why?

3. What is the geometric factor (aspect ratio) of the CA-M? Are there any limitations of its aspect ratio?

4. Is the electrical behavior of the CA-M the same when it works in the IDBC and in the EDBC? Why?

5. Why is the CA-M I_{DS} higher when it operates in the IDBC in comparison to the EDBC?

6. Is the longitudinal electric field near the drain region of CA-M higher when it operates in the IDBC compared to the EDBC? Justify.

7. In which drain bias configuration do the impact ionization and the ESD affect the CA-M the least? Why?

8. Why can we use the CA-MOSFET in the IDBC to reduce the drain parasitic capacitance, increase the unit voltage gain frequency (f_T), and also reduce the Miller capacitance effect in the amplifiers?

9. Describe the overlapping circular gate to be used in planar power MOSFETs (PPMs). What are the benefits of using it as a base-cell in the PPMs?

10. Why is the CA-M used as a hardness-by-design technique to improve the radiation tolerance of the MOSFETs?

CHAPTER 8

Wave Layout Style ("S" Gate Shape) for MOSFET

The wave layout style was specially proposed in order to overcome the aspect ratio (geometric factor), the total die area limitations, and the asymmetric geometry of the circular annular MOSFET (CA-M). Besides, this style intends to avoid the different electrical behaviors when it operates in the IDBC and EDBC, respectively [28].

Figure 8.1 illustrates a Wave MOSFET photograph (Figure 8.1.a), a layout (Figure 8.1.b) and a photograph of a planar power MOSFET implemented with a wave layout style (Figure 8.1.c), and a layout of a planar power MOSFET highlighting two unit cells with wave layout styles (n-type) (Figure 8.1.d). The latter is implemented by using the design rules of a commercial 350nm CMOS ICs technology of the ON Semiconductor via the MOSIS Educational Program, MEP-Research.

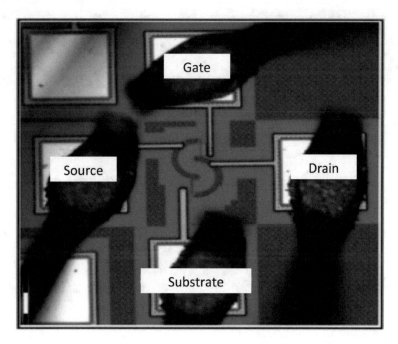

(a)

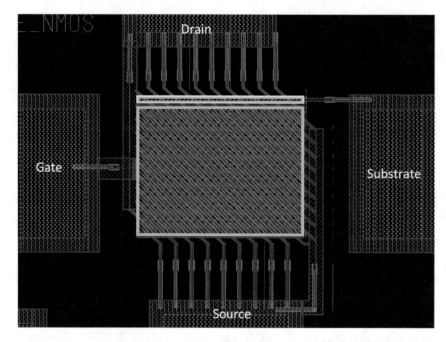

(b)

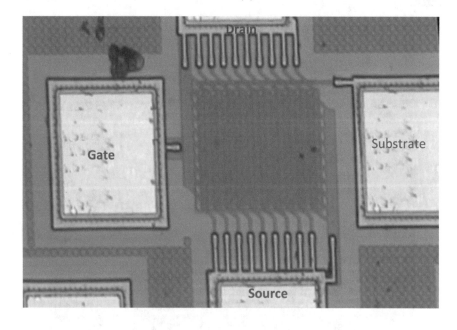

(c)

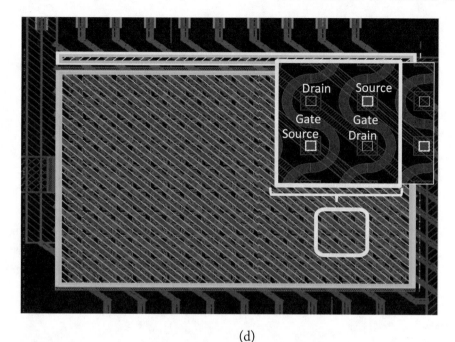

(d)

Figure 8.1: An example of a Wave MOSFET photograph (a), a layout (b), a photograph of a planar power MOSFET implemented with a wave layout style (c), and a layout of a planar power MOSFET highlighting two unit cells with wave layout styles (n-type) (d). The latter is implemented by using the design rules of a commercial 350 nm CMOS ICs technology of the ON Semiconductor via the MOSIS Educational Program, MEP-Research.

This innovative wave layout style was designed by separating the CA-MOSFET into two equal portions (semicircles) and then shifting one slice up to implement an S-shaped gate. Consequently, the circular annular MOSFET was converted into a symmetric geometrically device because the drain and source regions presented the same areas as the conventional rectangular SOI MOSFET. Besides, the wave layout style keeps the radial direction of IDS in each slice and it also presents the same aspect ratio [fg=$(W/L)_{conv.}$=2.π/ln(R2/R1), where R1 and R2, which are equal to R1 + L, are the initial and final radii that define the Wave MOSFET channel length (L)] of the CA-M [28].

The drain current vectors in the inferior and superior slices of the Wave MOSFET are illustrated in Figure 8.2 [28].

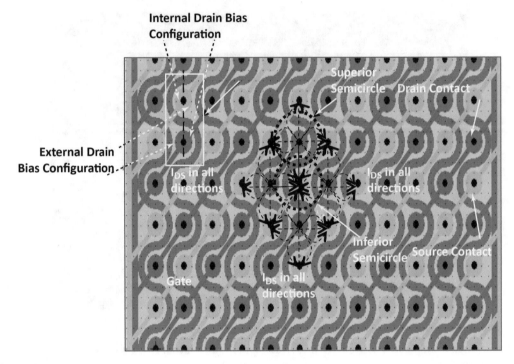

Figure 8.2: The IDS vectors in the Wave SOI nMOSFET.

We can observe in Figure 8.2 that the superior slice of the Wave MOSFET layout is con-figured to operate with internal (right site of the drain contact) and external (left site of the drain contact) drain bias simultaneously. The same occurs in the inferior semicircle of this structure, regarding internal and external source bias. Thus, this layout style is capable of reducing the internal and external drain/source bias effects, which occurs with the circular gate layout, operating only with one only type of drain bias configuration [28].

It is relevant to highlight that the Wave MOSFET drain current occurs in all radial directions. Subsequently, the systematic and random errors are critical due to the CMOS ICs manufacturing processes variations in the Wave MOSFET I_{DS} tending to be smaller than those found in the typical rectangular layout style as its drain current density occurs in a maximum of four directions [28]. Therefore, it can be an excellent alternative layout or hardness-by-design technique to boost the matching devices [28].

The absence of corners in the Wave MOSFET, where the boundaries are regulated by the reticulated pattern photo-masks, tends to further decrease the avalanche and Electrostatic Discharge (ESD) effects and increase its breakdown voltage in comparison to the standard rectangular MOSFET [28].

The wave layout style is capable of improving the integration factor (aspect ratio over total area ratio) of MOSFETs because it showed a better efficiency than Multifinger and Waffle layout styles (35.9% and 28.1%, respectively). The wave layout technique is also able to reduce the planar power SOI MOSFET size by 26.1% and 21.8% when compared to Multifinger and Waffle layouts, respectively [28].

Figure 8.3 illustrates the longitudinal electric field vectors near the edges of the Wave and standard rectangular layout styles (Bird's beak regions). Note that the longitudinal electrical field lines in the wave layout style are more distant from each other due to the radial direction near the source region in contrast to the rectangular counterpart. Consequently, its magnitude tends to be smaller than the one found in the classical rectangular layout. Knowing that the ionizing radiation effects in the MOSFETs are more significant in the regions where the electric field presents higher magnitudes, the wave layout style tends to be more ionizing radiation tolerant in relation to the classic rectangular layout style [29].

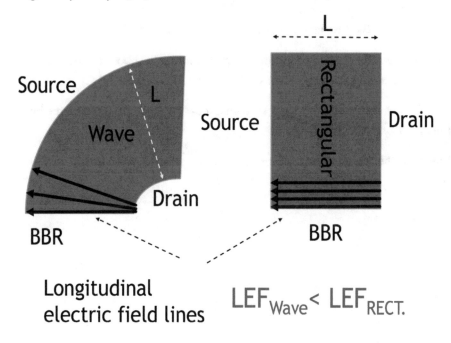

Figure 8.3: The longitudinal electric field vectors in the Bird's beak regions of the wave and rectangular layout styles.

In order to verify the Wave MOSFET tolerance in relation to the standard counterpart (same gate area), a study was performed in an ionizing radiation environment of 10 keV X-ray using a Shimadzu XRD-7000, with a cumulative dose up to 1.5 Mrad (from 500 krad to 1.5 Mrad with 500 krad steps) at a dose rate of 400 rad/s in unbiased devices. In this experiment, it was observed that the wave layout style is more ionizing radiation tolerant (TID) in comparison to the rectangular one in terms of threshold voltage, $I_{DS}/(W/L)$ as a function of V_{GS}, g_m, $I_{DS}/(W/L)$ as a function of V_{DS}, A_{V0}, and f_T after 1.5 Mrad TID effects, considering the same aspect ratio and bias conditions. Therefore, the Wave MOSFET can also be considered an alternative device to be used in CMOS ICs operating in space and medical applications [29].

PROBLEMS AND QUESTIONS

1. Why was the wave layout style for MOSFETs specially proposed?

2. What are the drain bias configurations of the superior and inferior slices of the Wave MOSFET?

3. What is the aspect ratio of the Wave MOSFET? Are there any limitations in its geometric factor as the circular annular MOSFET has? Justify.

4. Can the wave layout style be an alternative technique to boost the devices matching, ESD tolerance, and the breakdown voltage of the MOSFETs? Justify.

5. Can the Wave MOSFET be considered an alternative layout to improve the integration of planar power MOSFETs? Justify.

6. Why does the wave layout style present better radiation tolerance in relation to the classical rectangular MOSFET? Justify.

CHAPTER 9

Conclusions and Comments

This book summarizes some new layout techniques for MOSFETs that can be used and explored in order to boost the electrical performance and the ionizing radiation tolerance of the current and sophisticated planar CMOS ICs technology.

This innovative layout approach is able to incorporate new effects in the MOSFET structure, which is carefully described in this manuscript.

Regarding the simulations and experimental data results found so far, the author hopes to help the evolution and the progress of this important area of Electronic Engineering, mainly by combining low cost, high electrical performance, and ionizing radiation tolerance.

This is a new research field in micro/nanoelectronics and it is open to be further studied, investigated, explored, and used by students, researchers, CMOS ICs and devices (transistors, sensors, etc.), industries, and research institutes.

References

[1] Kilby, J. St. C. and R.R. Roup, "Transistor Amplifier Packaged in Steatite," *Electronics*, 1956. 1

[2] Kilby, J. St. C., "Invention of the Integrated Circuit," *IEEE Trans. on Electron Devices*, ED-23, 1976. 1

[3] Gordon E. Moore, "Cramming More Components onto Integrated Circuits," *Electronics*, vol. 38, no. 8, 1965. 1

[4] N. Damrongplasit, N. Xu, H. Takeuchi, R. J. Stephenson, N. W. Cody, A. Yiptong, X. Huang, Marek Hytha, R. J. Mears, and T.-J. K. Liu, "Comparative Study of Uniform Versus Supersteep Retrograde MOSFET Channel Doping and Implications for 6-T SRAM Yield," *IEEE Trans. on Electron Devices*, vol. 60, no. 5, pp. 1790–1793, May 1999. DOI: 10.1109/TED.2013.2253105. 1

[5] Y. Guo, X. Zhang, K. L. Low, K.-T. Lam, Y.-C. Yeo, and G. Liang, "Effect of Body Thickness on the Electrical Performance of Ballistic n-Channel GaSb Double-Gate Ultrathin-Body Transistor," *IEEE Trans. on Electron Devices*, vol. 62, no.3, pp. 788-794, 2015. DOI: 10.1109/TED.2014.2387194. 1

[6] S. Agarwal, T. B. Hook, M. Bajaj, K. McStay, W. Weike, and Yanting Zhang, "Transistor Matching and Fin Angle Variation in FinFET Technology," *IEEE Trans. on Electron Devices*, vol. 2, no. 4, pp. 1357-1359, 2015. DOI: 10.1109/TED.2015.2400221. 1

[7] C. B. Zota, L.-E. Wernersson, and E. Lind, "Multiple-Gate Field-Effect Transistors With Selectively Regrown Channels," *IEEE Trans. Electron Device Letters*, vol. 35, no. 3, pp. 342 - 344, 2014. DOI: 10.1109/LED.2014.2301843. 1

[8] K. Nayak, S. Agarwal, M. Bajaj, K. V. R. M. Murali, and V. R. Rao, "Random Dopant Fluctuation Induced Variability in Undoped Channel Si Gate all Around Nanowire n-MOSFET," *IEEE Trans. on Electron Devices*, vol. 62, no. 2, 2015. 1

[9] P. Chang; X. Liu, L. Zeng; K. Wei, and G. Du, "Investigation of Hole Mobility in Strained InSb Ultrathin Body pMOSFETs," *IEEE Trans. on Electron Devices*, vol. 62, no. 3, pp. 947–954, 2015. DOI: 10.1109/TED.2015.2388442. 1

[10] G. Hiblot, Q. Rafhay, F. Boeuf, and G. Ghibaudo, "Analytical Model for the Inversion Gate Capacitance of DG and UTBB MOSFETs at the Quantum Capacitance Limit,"

IEEE Trans. on Electron Devices, vol. 62, no. 5, pp. 1375–1382, 2015. DOI: 10.1109/TED.2015.2406116. 1

[11] G. Hiblot, Q. Rafhay, F. Boeuf, and G. Ghibaudo, "Analytical Model for the Inversion Gate Capacitance of DG and UTBB MOSFETs at the Quantum Capacitance Limit," *IEEE Trans. on Electron Devices*, vol. 62, no. 5, pp. 1375–1382, 2015. DOI: 10.1109/TED.2015.2406116. 1

[12] Wong, I., Chen, Y., Huang, S., Tu, W., Chen, Y., and Liu, C.W, "Junctionless Gate-All-Around pFETs Using In-situ Boron-Doped Ge Channel on Si," *IEEE Trans. Nanotechnology*, vol. 14, no. 5, pp. 878–882, 2015. DOI: 10.1109/TNANO.2015.2456182. 1

[13] J. Yi-Ruei; V. Thirunavukkarasu, C.-P. Wang, and Y.-C. Wu, "Performance Evaluation of Silicon and Germanium Ultrathin Body (1 nm) Junctionless Field-Effect Transistor With Ultrashort Gate Length (1 nm and 3 nm)," *IEEE Trans. Electron Device Letters*, vol. 36, no. 7, pp. 654–656, 2015. DOI: 10.1109/LED.2015.2437715. 1

[14] A. Villalon, G. Le Carval, S. Martinie, C. Le Royer, M.-A. Jaud, and S. Cristoloveanu, "Further Insights in TFET Operation," *IEEE Trans. on Electron Devices*, vol. 61, no. 8, pp. 2893–2898, 2014. DOI: 10.1109/TED.2014.2325600. 1

[15] S. K. Mohapatra, K. P. Pradhan, D. Singh, and Sahu, P.K., "The Role of Geometry Parameters and Fin Aspect Ratio of Sub-20nm SOI-FinFET: An Analysis Towards Analog and RF Circuit Design," *IEEE Trans. Nanotechnology*, vol. 14, no. 3, pp. 546–554, 2015. DOI: 10.1109/TNANO.2015.2415555. 1

[16] Y. Ko, P. Roblin, A. Zarate-de Landa, J. A. Reynoso-Hernandez, D. Nobbe, C. Olson, and F. J. Martinez, "Artificial Neural Network Model of SOS-MOSFETs Based on Dynamic Large-Signal Measurements," *IEEE Trans. Microwave Theory and Techniques*, vol. 62, no. 3, pp. 491–501, 2014. DOI: 10.1109/TMTT.2014.2298372. 1

[17] E. D. Litta, P.-E. Hellstrom, and M. Ostling, "Enhanced Channel Mobility at Sub-nm EOT by Integration of a TmSiO Interfacial Layer in HfO2/TiN High-k/Metal Gate MOSFETs," *IEEE Journal of the Electron Devices Society*, vol. 3, no. 5, pp. 397-404, 2015. DOI: 10.1109/JEDS.2015.2443172. 1

[18] S. Tewari, A. Biswas, and A. Mallik, "Investigation on High-Performance CMOS With p-Ge and n-InGaAs MOSFETs for Logic Applications," *IEEE Trans. Nanotechnology*, vol. 14, no. 2, pp. 275–281, 2015. DOI: 10.1109/TNANO.2015.2390295. 1

[19] I. Wong, Y. Chen, S. Huang, W. Tu, Y. Chen, and C.W. Liu, "Junctionless Gate-All-Around pFETs Using In-situ Boron-Doped Ge Channel on Si," *IEEE Trans. Nanotechnology*, vol. 14, no. 5, pp. 878–882, 2015. DOI: 10.1109/TNANO.2015.2456182. 1

[20] F. Bashir, S. A. Loan, M. Rafat, A. R. M. Alamoud, and S. A. Abbasi, "A High-Performance Source Engineered Charge Plasma-Based Schottky MOSFET on SOI," *IEEE Trans. Electron Devices*, vol. 62, no. 10, pp. 3357-3364, 2015. DOI: 10.1109/TED.2015.2464112. 1

[21] Y. Kuang, R. Huang, Y. Tang, W. Ding, L. Zhang, and Y. Wang, "Flexible Single-Component-Polymer Resistive Memory for Ultrafast and Highly Compatible Nonvolatile Memory Applications," *IEEE Trans. Electron Device Letters*, vol. 31, no. 7, pp. 758–760, 2010. DOI: 10.1109/LED.2010.2048297. 1

[22] S. P. Gimenez, R. M. G. Ferreira, and J. A. Martino, "Early Voltage Behavior in Circular Gate SOI nMOSFET using 0.13 μm Partially-Depleted SOI CMOS Technology," *ECS Transactions*, vol. 4, no. 1, pp. 309–318, 2006. 1, 2, 6, 7, 8, 50

[23] M. A. G. Silveira, K. H. Cirne, J. A. Lima, L. E. Seixas, N. H. Medina, M. H. Tabacniks, N. Added, S. P. Gimenez, W. Melo, and M. B. Dupret, "Performance of Electronics Devices Submitted to X-ray and High Energy Proton Beams," *Nuclear Instruments & Methods in Physics Research. Section B, Beam Interactions with Materials and Atoms*, vol. 11, p. 702–707, 2011. 1, 2, 8, 50

[24] K. H. Cirne, J. A. Lima, L. E. Seixas, M. A. G. Silveira, M. Barbosa, M. Tabacnicks, N. Added, S. P. Gimenez, and N. Medina, W., "Comparative Study of the Proton Beam Effects between the Conventional and Circular Gate MOSFETs," *Nuclear Instruments & Methods in Physics Research. Section B, Beam Interactions with Materials and Atoms*, vol. 11, p. 688–691, 2011. 1, 2, 8, 50

[25] J. A. Lima and S. P. Gimenez, "A Novel Overlapping Circular-Gate Transistor and its Application to Power MOSFETs," *ECS Transactions*, vol. 23, no. 1, pp. 361–369, 2009. 1, 2, 8, 48, 50

[26] Lima, J. A., S. P. Gimenez, and K. H. Cirne, "Modeling and Characterization of Overlapping Circular-Gate MOSFET and Its Application to Power Devices," *IEEE Transactions on Power Electronics*, v. 27, p. 1622–1631, 2012. DOI: 10.1109/TPEL.2011.2117443. 1, 2, 8, 48, 50

[27] S. P. Gimenez (Inventor) and FEI University Center (Titular), Wave gate ("S" shaped gate) MOSFET, patent no. PI0802634-3 A2, submission data: 07/08/2008, publication date: 30/03/2010 (RPI 2047). 1, 2, 8

[28] S. P. Gimenez, "The Wave SOI MOSFET: A New Accuracy Transistor Layout to Improve Drain Current and Reduce Die Area for Current Drivers Applications," *ECS Transac-

tions, vol. 19, no. 4, pp. 153–158, 2009. DOI: 10.1149/1.3117404. 1, 2, 12, 44, 53, 55, 56, 57

[29] R. N. Souza, M. A. G. Silveira, and S. P. Gimenez, "Mitigating MOSFET Radiation Effects by Using the Wave Layout in Analog ICs Applications," *Journal of Integrated Circuits and Systems*, vol. 10, p. 30–37, 2015. 1, 2, 8, 12, 44, 57, 58

[30] S. P. Gimenez (Inventor) and FEI University Center (Titular), Hexagonal gate (Diamond) MOSFET, patent no. PI0802745-5 A2, submission data: 07/08/2008, publication date: 23/03/2010 (RPI 2046). 1, 2, 8, 10, 11, 12

[31] S. P. Gimenez, "Diamond MOSFET: An innovative layout to improve performance of ICs," *Solid-State Electronics*, vol. 54, pp. 1690–1699, 2010. DOI: 10.1016/j.sse.2010.08.011. 1, 2, 8, 10, 11, 12

[32] S. P. Gimenez, D. M. Alati, M. A. G. Silveira, L. E. Seixas, W. R. Mello, N. Added, N. Medina, and M. H. Tabacknics, "Improving Proton Radiation-Robustness of Integrated Circuits by Using Diamond Layout Style," In: *Radecs 2012 - Radiation and Its Effects on Components and Systems*, 2012. 1, 2, 8, 10, 11, 12, 18

[33] S. P. Gimenez, R. D. Leoni, C. Renaux, and Flandre, D., "Using diamond layout style to boost MOSFET frequency response of analogue IC," *Electronics Letters*, vol. 50, pp. 398–400, 2014. DOI: 10.1049/el.2013.4038. 1, 2, 8, 10, 11, 12

[34] S. P. Gimenez, E. D. Neto, V. V. Peruzzi, C. Renaux, and D. Flandre, "A compact Diamond MOSFET model accounting for the PAMDLE applicable down the 150 nm node," *Electronics Letters*, vol. 50, pp. 1618–1620, 2014. DOI: 10.1049/el.2014.1229. 1, 2, 8, 10, 11, 12, 13

[35] S. P. Gimenez, E. H. S. Galembech, C. Renaux, and D. Flandre, "Diamond layout style impact on SOI MOSFET in high temperature environment," *Microelectronics and Reliability*, vol. 55, pp. 783–788, 2015. DOI: 10.1016/j.microel.2015.02.015. 1, 2, 8, 10, 11, 12, 13

[36] S. P. Gimenez, E. H. S. Galembeck, C. Renaux, and D. Flandre, "Impact of Using the Octagonal Layout for SOI MOSFETs in High Temperature Environment," *IEEE Trans. on Device and Materials Reliability (Letters)*, vol. 99, p. 1–1, 2015. DOI: 10.1109/tdmr.2015.2474739. 1, 2, 8, 10, 11, 12, 13

[37] Silva, G. A. and Gimenez, P. S., "Boosting the performance of the planar power MOSFET By using Diamond layout style," In: *29th Symposium on Microelectronics Technology and Devices (SBMicro)*, vol. 1, no. 1, pp. 1–4, 2014. DOI: 10.1109/sbmicro.2014.6940107. 1, 2, 8, 10, 11, 12, 13, 14

[38] Neto, E. D. and Gimenez, S. P., "Applying the Diamond Layout Style for FinFET," *ECS Transactions*, vol. 49, no. 1, pp. 535–542, 2012. DOI: 10.1149/04901.0535ecst. 1, 2, 8, 10, 11, 12, 14, 15

[39] S. P. Gimenez, and D. M. Alati, "Electrical behavior of the diamond layout style for MOS-FETs in X-rays ionizing radiation environments," *Microelectronic Engineering*, vol. 148, p. 85–90, 2015. DOI: 10.1016/j.mee.2015.09.001. 1, 2, 8, 10, 11, 12, 18

[40] L. E. Seixas, M. A. G. Silveira, N. Medina., V. A. P Aguiar, N. Added, S. and P. Gimenez," A New Test Environment Approach to SEE Detection in MOSFETs," *Advanced Materials Research*, vol. 1083, p. 197–201, 2015. DOI: 10.4028/www.scientific.net/AMR.1083.197. 1, 2, 8, 10, 11, 12, 15

[41] S. P. Gimenez (Inventor) and FEI University Center (Titular), Multi-Edges MOSFETs, patent no. PI0903005-0 A2, submission data: 28/08/2009, publication date: 10/05/2011 (RPI 2105). 1, 2, 8, 21, 22

[42] S. P. Gimenez, and D. M. Alati, "OCTO SOI MOSFET: An Evolution of the Diamond to Be Used in the Analog Integrated Circuits," In: *EUROSOI*, Granada, Spain, 2011. 1, 2, 8, 21, 25

[43] L. N. S. Fino, C. Renaux, D. Flandre, and S. P. Gimenez, "Experimental Study of the OCTO SOI nMOSFET to Improve the Device Performance," In: *EUROSOI*, Montpellier, France, 2012. 1, 2, 8, 21, 22, 26

[44] L. N. S. Fino, C. Renaux, D. Flandre, and S. P. Gimenez, "Improving Unit Voltage Gain Frequency of Integrated Circuits by Using OCTO Layout Style," In: *EUROSOI*, Paris, France, 2013. 1, 2, 8, 21, 22, 26

[45] L. N. S. Fino, M. A. G. Silveira, C. Renaux, D. Flandre, and S. P. Gimenez, "Total Ionizing Dose Effects on the Digital Performance of Irradiated OCTO and Conventional Fully Depleted SOI MOSFET," In: *RADECS*, Oxford, England, 2013. 1, 2, 8, 21, 22, 27

[46] L. N. S. Fino, M. A. G. Silveira, C. Renaux, D. Flandre, and S. P. Gimenez, "Improving the X-Ray Radiation Tolerance of the Analog ICs by Using OCTO Layout Style," In: *28th Symposium on Microelectronics Technology and Devices (SBMicro)*, Paraná, Brazil, 2013. DOI: 10.1109/SBMicro.2013.6676166. 1, 2, 8, 21, 22, 27, 28

[47] L. N. S. Fino, M. A. G. Silveira, C. Renaux, D. Flandre, and S. P. Gimenez, "Boosting the Radiation Hardness and Higher Reestablishing Pre-Rad Conditions by Using OCTO Layout Style for MOSFETs," In: *29th Symposium on Microelectronics Technology and Devices (SBMicro)*, Sergipe, Brazil, 2014. 1, 2, 8, 21, 22, 27, 28

[48] L. N. S. Fino, M. A. G. Silveira, C. Renaux, D. Flandre, and S. P. Gimenez, "The Influence of Back Gate Bias on the OCTO SOI MOSFETs Response to X-ray Radiation," *Journal of Integrated Circuits and Systems* (JICS), vol. 10, pp. 43–48, 2015. 1, 2, 8, 21, 22, 27, 28

[49] L. N. S. Fino, M. A. G. Silveira, C. Renaux, D. Flandre, and S. P. Gimenez, "Boosting the total ionizing dose tolerance of digital switches by using OCTO SOI MOSFET," *Semiconductor Science and Technology*, vol. 30, pp. 105024–12p, 2015. DOI: 10.1088/0268-1242/30/10/105024. 1, 2, 8, 21, 22, 23, 24, 25, 27, 28

[50] S. P. Gimenez (Inventor) and FEI University Center (Titular), Circular Eyes, patent no. PI0905289-5 A2, submission data: 02/12/2009, publication date: 19/07/2011 (RPI 2115). 1, 2, 8, 28, 31, 33

[51] S. P. Gimenez, M. M. Correia, E. D. Neto, and C. Silva," An innovative Ellipsoidal layout style to further boost the electrical performance of MOSFETs," *IEEE Trans. Electron Device Letters*, v. 36, no. 7, pp. 705–706, 2015. DOI: 10.1109/LED.2015.2437716. 1, 2, 8, 31, 33, 35, 37, 38

[52] M. M. Correia and S. P. Gimenez, "Boosting the electrical performance of MOSFET switches by applying Ellipsoidal layout style," In: *30th Symposium on Microelectronics Technology and Devices (SBMicro)*, pp. 1–4, 2015. DOI: 10.1109/sbmicro.2015.7298122. 1, 2, 8, 31, 33, 37, 38

[53] Gimenez, S. P., and Alati, D. M., "FISH SOI MOSFET: An Evolution of the Diamond SOI Transistor for Digital ICs Applications," *ECS Transactions*, vol. 35, no. 5, pp. 163-168, 2011. DOI: 10.1149/1.3570792. 1, 2, 8, 41, 43, 45

[54] S. P. Gimenez, D. M. Alati, E. Simoen, and C. Claeys, "FISH SOI MOSFET: Modeling, Characterization and Its Application to Improve the Performance of Analog ICs," *Journal of the Electrochemical Society*, v. 158, p. H1258–H1264, 2011. DOI: 10.1149/2.091112jes. 1, 2, 8, 41, 43, 45

[55] X. Hou, F. Zhou, R. Huang, and Xing Zhang, "Corner Effects in Vertical MOSFETs," In: *7th International Conference on Solid-State and Integrated Circuits Technology*, vol. 1, pp. 134–137, 2004. 5, 6

[56] D. R. Oliveira and S. P. Gimenez, "Using Cynthia SOI MOSFET to Improve Voltage Gain of Analog Integrated Circuits," *ECS Transactions*, vol. 23, no. 1, pp. 381–388, 2009. DOI: 10.1149/1.3183742. 5, 6, 9

[57] J. P. Colinge and C. A. Colinge, "*Phisics of Semiconductor Devices*," Kluwer Academic Publishers, 2002., 9

[58] B. Razavi, *Design of Analog CMOS Integrated Circuits*, McGraw Hill, 2001. 9, 10, 43

[59] Y. V. Bogatyrev, S. B. Lastovskiy, F.P., Korshunov, V. I. Kulgachev, A. S. Turtsevich, S. V. Shwedov, V. S. Malyshev, and A. M. Yarmolik, "Radiation Effects in Elements of Submicron CMOS Integrated Circuits With Various Kinds of Isolation," In: *Microwave and Telecommunication Technology (CriMiCo)*, 2013 23rd International Crimean Conference, pp. 922–923, 2013. 16, 18

[60] H. J. Barnaby, M. L. McLain, I. S. Esqueda, and X. J. Chen, "Modeling Ionizing Radiation Effects in Solid State Materials and CMOS Devices," *IEEE Transactions on Circuits and Systems*, vol. 56, no. 8, 2009. DOI: 10.1109/tcsi.2009.2028411. 16, 18

[61] B.-Y. Chou, C.-S. Lee, C.-L.Yang, W.-C. Hsu, H.-Y. Liu, M.-H. Chiang, W.-C. Sun, S.-Y. Wei, and S.-M. Yu, "TiO2-Dielectric AlGaN/GaN/Si Metal-Oxide-Semiconductor High Electron Mobility Transistors by Using Nonvacuum Ultrasonic Spray Pyrolysis Deposition," *IEEE Electron Devices Letters*, vol. 35, no. 11, pp. 1091–1093, 2014. DOI: 10.1109/LED.2014.2354643.

[62] R. Minixhofer, N. Feilchenfeld, M. Knaipp, G. Röhrer, J.M. Park, M. Zierak, H. Enichlmair, M. Levy, B. Loeffler, D. Hershberger, F. Unterleitner, M. Gautsch, K. Chatty, Y. Shi, W. Posch, E. Seebacher, M. Schrems, J. Dunn+, and D. Harame, "A 120V 180nm High Voltage CMOS smart power technology for System-on-chip integration," In: *Proceedings of The 22nd International Symposium on Power Semiconductor Devices & ICs*, Hiroshima, pp. 75–78, 2010.

[63] X. Wang, X. Guan, S. Fan, H. Tang, H. Zhao, L. Lin, Q. Fang, J. Liu, A. Wang, and L. Yang, "ESD-Protected Power Amplifier Design in CMOS for Highly Reliable RF ICs," *IEEE Transactions on Industrial Electronics*, vol. 58, no. 7, pp. 2736–2743, 2011. DOI: 10.1109/TIE.2010.2057234.

[64] M-D. Ker and C-T. Yeh, "On the Design of Power-Rail ESD Clamp Circuits With Gate Leakage Consideration in Nanoscale CMOS Technology," *IEEE Transactions on Devices and Materials Reliability*, vol. 14, no. 1, pp. 536–544, 2014. DOI: 10.1109/TDMR.2013.2280044.

[65] Z. Wang, R.-C. Sun, J. J. Liou, and D.-G.Liu, "Optimized pMOS-Triggered Bidirectional SCR for Low-Voltage ESD Protection Applications," *IEEE Transactions on Electron Devices*, vol. 61, no. 7, pp. 2588–2594, 2014. DOI: 10.1109/TED.2014.2320827.

[66] Kosuge, Y. and Matsuzaki, T., "The optimum gate shape and threshold for target tracking," *SICE Annual Conf.*, vol.2, p. 2152, 2157, Aug. 2003. 33

[67] Y. Li and C.-H. Hwang, "The effect of the geometry aspect ratio on the silicon ellipse-shaped surrounding-gate field-effect transistor and circuit," *Semiconductor Science Technology*, vol. 24, no. 9, pp. 095018-1–095018-8, 2009. DOI: 10.1088/0268-1242/24/9/095018. 53

[68] L. Zhang et al., "Modeling short-channel effect of elliptical gate-all-around MOSFET by effective radius," *IEEE Electron Device Letters*, vol. 32, no. 9, pp. 1188–1190, Sep. 2011. DOI: 10.1109/LED.2011.2159358. 33

[69] S. Kumar and S. Jha, "Impact of elliptical cross-section on the propagation delay of multi-channel gate-all-around MOSFET based inverters," *Microelectronics Journal*, vol. 44, no. 9, pp. 844–851, 2013. DOI: 10.1016/j.mejo.2013.06.003. 33

[70] Jean-Pierre Colinge, *Silicon-On-Insulator Technology: Materials to VLSI*, 3rd ed., Kluwer Academic Publishers, 2004. 47, 48

[71] Dantas, L. P. and Gimenez, S. P., "Comparison Between Harmonic Distortion in Circular Gate and Conventional SOI nMOSFET Using 0.13 µm Partially-Depleted SOI CMOS Technology Electrical Characterization," *ECS Transactions*, vol. 11, no. 3, pp. 85–96, 2007. 48, 49

[72] Cirne, K. H., Silveira M. A. G., Lima J. A., Seixas, L. E., and Gimenez S. P., "X-Ray Radiation Effects in the Circular-Gate Transistors," *ECS Transactions*, vol. 35, no. 5, pp. 259–264, 2011. DOI: 10.1149/1.3570804. 50

About the Author

Salvador Pinillos Gimenez was born in São Paulo, Brazil, in 1962. He received a graduate degree in Electrical Engineering from UMC University, São Paulo, Brazil, in 1984. He earned his M.S. (Microelectronics Laboratory, LME) and Ph.D. (Integrated Systems Laboratory, LSI) in Electrical Engineering from the University of São Paulo, Brazil, in 1990 and 2004, respectively. He worked as a Product Engineer at Dimep S.A (1987–1992), and Tracecom Telecommunications Systems S.A (1993). From 1994 to 1999, he worked at Ford (Electronic Division, after Visteon) in Guarulhos, São Paulo, Brazil, as a Component Engineer and Supply Quality Assurance and then Quality & Productivity Coordinator. Since 1999, he has been a professor and researcher at FEI University Center, Brazil, and became a full professor in 2010. He has authored textbooks on Microcontrollers and holds industrial patents on MOSFET with innovative gate geometries. His major fields of study include innovative non-standard MOSFETs structures, analog and digital CMOS ICs designs, and evolutionary electronic by developing analog CMOS ICs tools. Dr. Gimenez is a member of the Microelectronic Brazilian Society (SBMicro) and IEEE member (M'13). Since October 2015, he has participated on the editorial board (Associate Editor) of the *Electronics Letters Journal* (Institution of Engineering and Technology, IET - England, Wales, and Scotland).

Printed in the United States
by Baker & Taylor Publisher Services